Hypothesis Testing

The Ultimate Beginner's Guide to Statistical Significance

by Arthur Taff

TEACHING NERDS

Table of Contents

Introduction

When you are testing for statistical significance, the equation you are asking is, how unlikely would this outcome have been if it just happened by random chance?

For instance, say you are a farmer trying out a new type of pig food. You are trying to determine if the new food makes the pigs gain weight faster. To do this test, you try the new food on 10 pigs, and the old food on 10 pigs. After a month, you measure the weight of all the pigs and find that the pigs who ate the new food gained more weight. Does that mean the new food caused the increased weight gain?

Well, maybe. Maybe the new food caused that result, or maybe you just happened to give the new food to the 10 pigs that were always going to gain the most weight no matter what. How can you tell?

What we are doing with statistical significance calculations is determining how unlikely an outcome was to occur by random chance, and then deciding if that probability is unlikely enough that we can conclude something other than random chance caused that outcome.

The two most important things in a statistical significance calculation are the distance the average value of your measured data is from what you are comparing it against, and the standard deviation of what you are measuring. It is easy to understand how the difference in measurements is important.

If I am making some measurements, and I measure a difference of 100 compared to the typical mean value, that is more likely to be significant than if I measured a difference of 5. The greater the difference in measurements, the greater the significance, assuming that all other values are equal.

Standard Deviation

Standard deviation is the second important topic in calculating statistical significance. It is worth going over how the standard deviation works. Standard deviation is a way of measuring how spread out your measured values are. If the values you are measuring are all clustered together, they will have a low standard deviation. If they have a lot of variation in the measurements, they will have a high standard deviation.

As an example, pick a coin from your pocket, for instance, a quarter if you are in the United States. If you measure the weight of 10 of those coins, they will probably all weigh about the same.

After all, they were all manufactured to be similar. Now if you go outside and find 10 rocks approximately the size of those coins and measure the weight of those rocks, there will be a lot more variation in the weight of the rocks. The standard deviation of the weight of the rocks is higher than it is for the coins.

The image below shows what we typically think of when we think about standard deviation. There is a mean value at the center of a normal curve. If you make another measurement of the same type of data it will fall somewhere on that normal curve, with decreasing likelihood the farther you get away from the mean value.

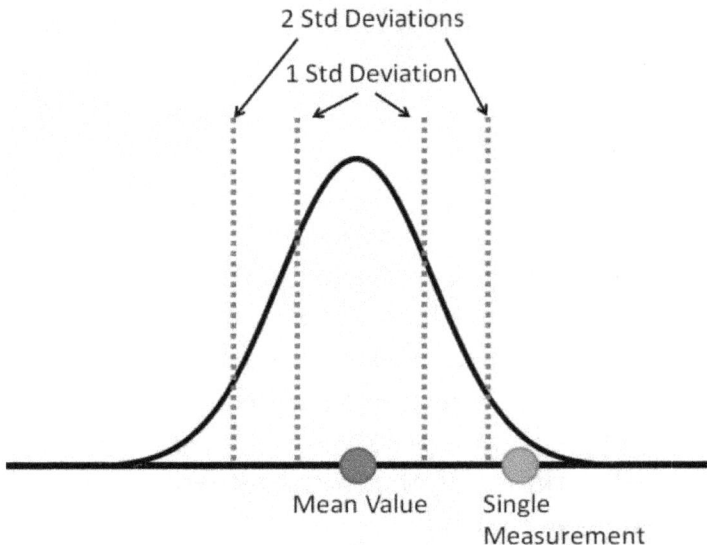

With a typical normal curve:

- 68 percent of the data will fall within 1 standard deviation of the mean
- 95 percent of the data be within 2 standard deviations
- 99.7 percent of the data is within 3 standard deviations

With hypothesis testing, what we are doing is turning that chart around and asking the question in reverse. Now what we are doing is putting a normal curve around our measured data, and asking the question "How likely is it that this measured data comes from the same source as the reference data?" We are asking how many standard deviations the reference value is from our mean value. This is shown in the chart below.

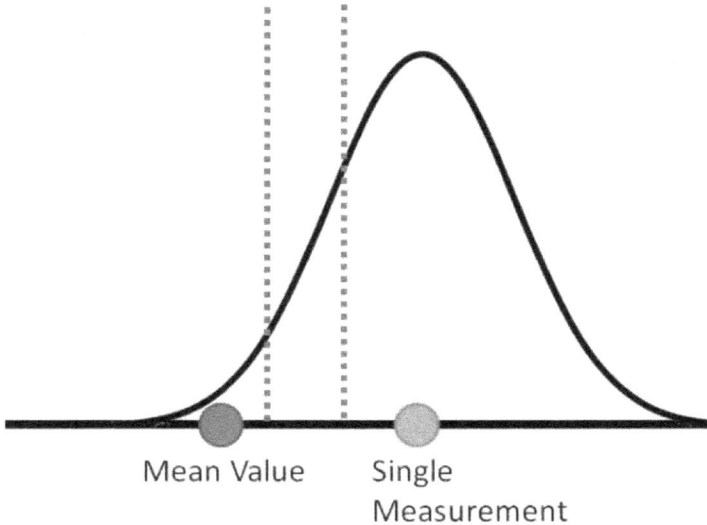

Mean Value   Single
              Measurement

Now this might seem like it was a useless distinction. If the reference value was two standard deviations away from the measured value, then the measured value will be two standard deviations away from the reference value. That is completely true, but only if we only have a single piece of measured data.

Chapter 1: You Need to Understand This

We now get to the single most important concept in understanding statistical significance. If you understand this, then you understand statistical significance. The rest of it is just knowing when to apply which equation. Because of its importance, we will spend the next couple pages going over this concept a few difference ways.

The concept is this: **We do not care about the standard deviation of our data. What we care about is the standard deviation of the average value of all of our measurements. And that standard deviation of the average can change as you make additional measurements.**

This is shown in the chart below. Like the chart above, it has a normal curve centered on the mean values of the measured data. However, due to the fact that there are more measurements in this data, rather than just the single measurement in the chart above, this normal curve is narrower

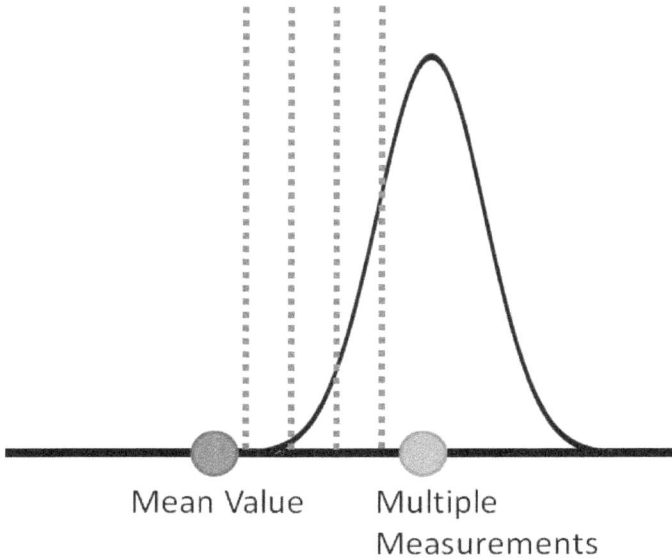

Mean Value   Multiple Measurements

Since the normal curve is narrower, the reference value falls farther away from the measured average, in terms of the number of standard deviations. As a result, we will conclude that it is less likely that our measured values and the reference value came from the same set of data.

Why Does the Standard Deviation of The Mean Decrease with More Measurements?

We stated that what we are interested in is the standard deviation of the average value of all of the measurements we make. The standard deviation of that average value decreases as you increase the number of measurements made. Why is that?

Fortunately, we don't have to use complicated math to explain this topic. You have almost certainly seen it before, although you probably didn't think about it in these terms.

The example we will use to demonstrate what happens is the result you get when you roll 1 die, vs 2 dice vs. an increasing number of dice.

If you roll a single die, you are equally likely to get a 1, 2, 3, 4, 5, or 6. Each value will come up one-sixth of the time, so the probability distribution of a single die rolled six times will have each value coming up one time.

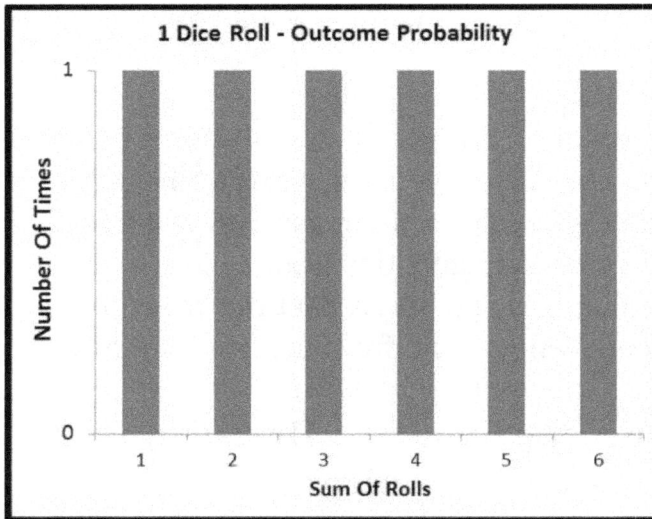

Now what happens if you roll two dice, and add their values? Well, there are 36 different permutations of die rolls that you can get out of two dice, 6 values from the first die, multiplied by 6 values from the second die. However, there are only 11 different sums that you can get from those two dice, the values of 2 through 12.

Those 36 different die permutations don't map evenly onto the 11 different possible sums. You are more likely to get values in the middle of the range than values on the edges. The probability distribution of the sum of two dice is shown below.

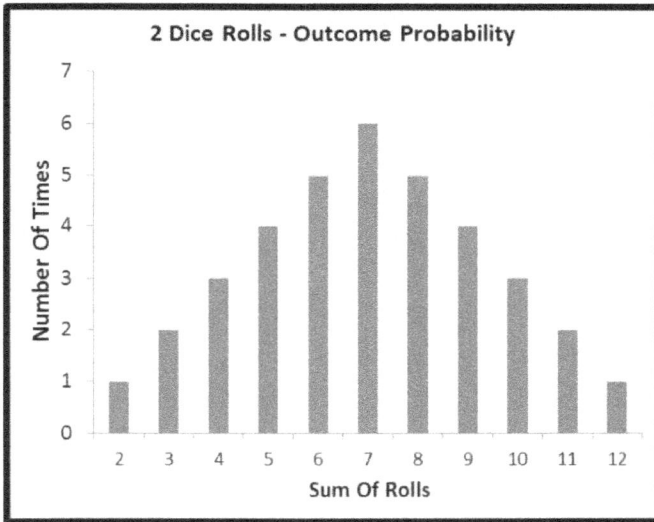

2 Dice Rolls - Outcome Probability

The single most likely sum is a 7, which is why casinos make that value a number that they win on in the game of craps.

For our purposes, the key point is that the probability of outcomes is more concentrated in the center of the graph for two die rolls than it is for a single die roll. That concentration of probability into the center of the graph doesn't stop with 2 dice. If you sum the value of 3 dice there are 216 different permutations of rolls (6 * 6 * 6) mapped onto 16 possible values (3-18). The probability distribution for 3 dice summed is shown below.

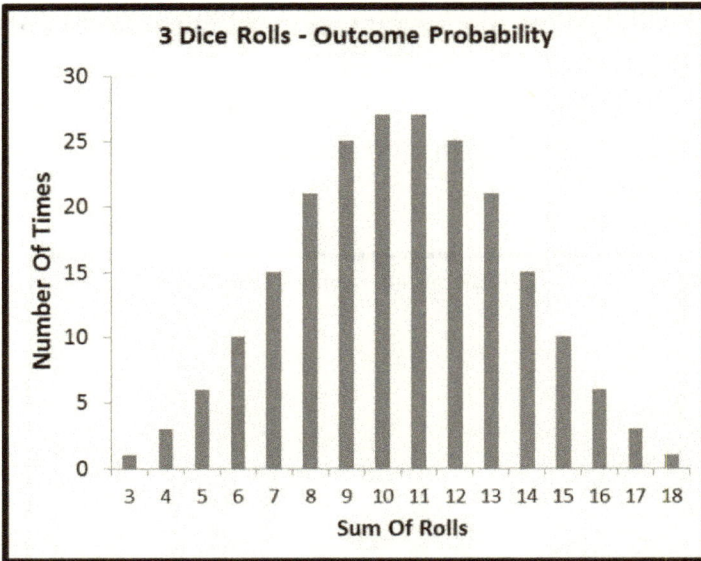

3 Dice Rolls - Outcome Probability

Even though it isn't quite as visually obvious as going from 1 die to 2 dice, summing 3 dice has a greater concentration of probability in the center of the graph than the sum of 2 dice. That process would continue if we kept rolling additional dice.

So far with these dice, we've talked about the sum of the dice values. Now let's talk about the average value. To calculate the average value, just take the sum and divide by the number of dice. So, for instance if you rolled a sum of 7 on 2 dice, then your average value was 3.5.

The probability distribution of the average value for 1, 2, and 3 dice rolled is shown in the plot below.

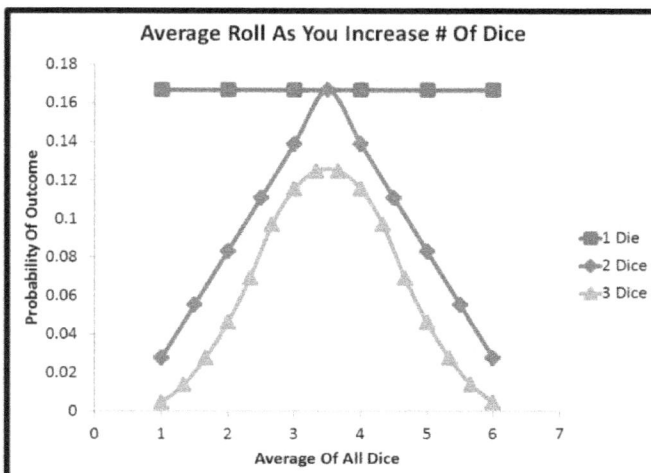

Average Roll As You Increase # Of Dice

This plot makes a few things obvious:

- Not matter how many dice are rolled, the average value is always centered on 3.5. This is because the average of all the numbers on a single die, (average of 1, 2, 3, 4, 5, 6) is 3.5

- As you increase the number of dice, the probability of getting an average value that is on the edges of the range decreases dramatically. The bulk of the probability shifts to the center of the graph.

It might be surprising that the probability for every single possible average for 3 dice is lower than their counterpart for 2 dice, and also for 1 die. That is because as you increase the number of dice, the probability is spread among more possible outcomes. There are only 6 possible outcomes with 1 die, but 11 possible outcomes with 2 die and 16 possible outcomes with 3 die.

13

In order to get consistent probability distributions with different numbers of dice, we can use a histogram and 'binning' to make sure the probability is spread among the same number of outcomes.

That wouldn't plot very well for 1, 2 and 3 dice; however here is a binned result of the probability distribution for the average roll when rolling 5, 10, and 20 dice.

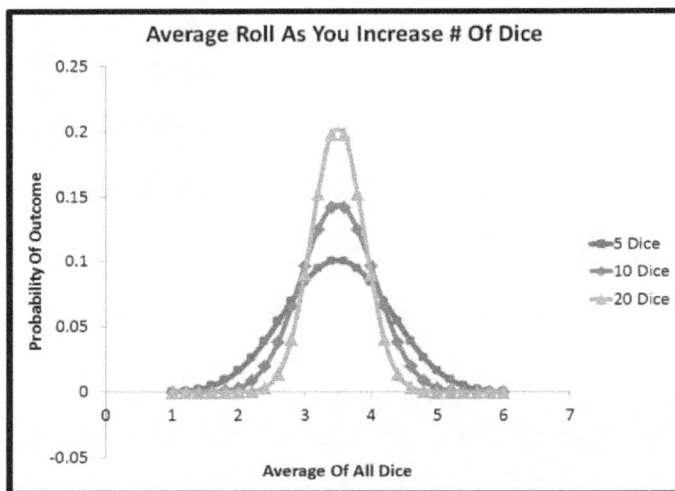

As you can see, the greater the number of rolls, the more the probability distribution of the average value is clustered around the average value of 3.5 and less at the edges. This means that the standard deviation of the average is decreasing as you increase the number of samples used to calculate it.

We can, in fact, calculate the standard deviation of the average for a given number of dice. For a single die, this is just the population standard deviation of [1, 2, 3, 4, 5, 6], which is 1.7078.

For two dice, it would be the population standard deviation of the 36 possible average values (11 distinct values) of [1, 1.5, 1.5, 2, 2, 2, 2.5, 2.5, 2.5, 2.5, 3, 3, 3, 3, 3, 3.5, 3.5, 3.5, 3.5, 3.5, 3.5, 4, 4, 4, 4, 4, 4.5, 4.5, 4.5, 4.5, 5, 5, 5, 5.5, 5.5, 6], which is 1.207615.

Fortunately, there are more efficient ways to calculate the standard deviation than listing out every single value (by using a weighted average of the squared difference from the mean). When you calculate the standard deviation of the average for all numbers of dice between 1 and 20, you get the plot below.

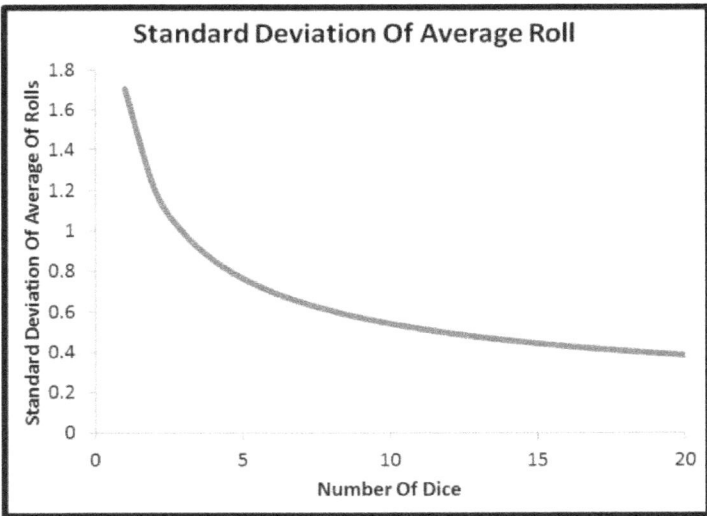

As expected, we see that the standard deviation of the average continues to drop as we increase the number of samples. The other notable thing about this plot is that the rate of change begins to level off. Adding a few more dice drastically decreased the standard deviation of the average at the beginning, but it takes a greater and greater number of dice to get the same amount of change.

In practical terms, what this means for statistical significance is that there is a diminishing return to getting more data to use in your statistical significance calculation. At the beginning additional data will make a large change, however, eventually the cost of acquiring additional data will outweigh the benefit.

The plot below is the same as the plot above, except with a regression curve fit on the data.

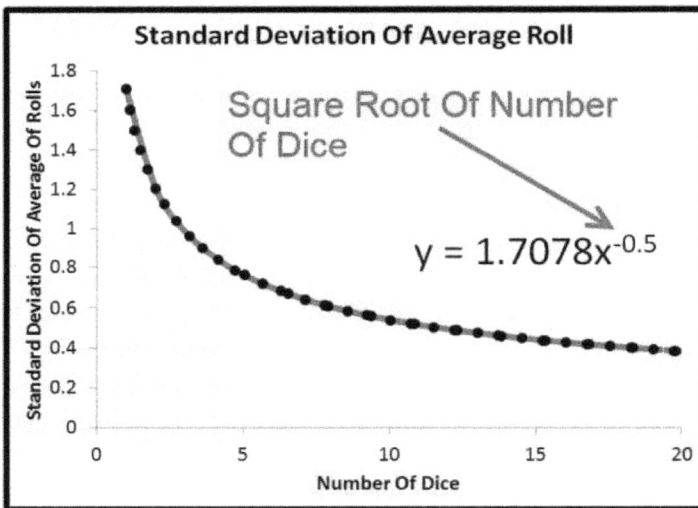

Standard Deviation Of Average Roll

Square Root Of Number Of Dice

$y = 1.7078x^{-0.5}$

Y-axis: Standard Deviation Of Average Of Rolls (0, 0.2, 0.4, 0.6, 0.8, 1, 1.2, 1.4, 1.6, 1.8)

X-axis: Number Of Dice (0, 5, 10, 15, 20)

What we can see is that the regression fit the data exactly. The rate of change of the standard deviation of the average is equal to the standard deviation of the population, multiplied by the number of data points raised to the power of negative one-half. A power of one-half is the same as a square root. A negative power is the same as dividing by that number. Rewriting the equation to incorporate those changes

$$y = \frac{1.7078}{\sqrt{x}}$$

Here the 1.7078 is the standard deviation of the population of average values with 1 sample (i.e. [1, 2, 3, 4, 5, 6]). We will denote that with a sigma. Here 'x' is the number of dice. In later problems, instead of 'x' denoting the number of dice, the equations use 'n' to denote the number of measurements. If we use those symbols, this equation becomes

$$y = \frac{\sigma}{\sqrt{n}}$$

Although it varies slightly depending on the problem in question, that sigma and the square root of n appear in pretty much every variation of the statistical significance equations that we use in this book. What they are demonstrating is that the standard deviation of the average of the samples is the standard deviation of the samples (sigma) divided by the square root of the number of samples.

Or to put it another way, as you increase the number of samples, the resulting average of your measurements is increasingly likely to be close to the true population average, just like we saw with the dice rolling.

Read This FIRST - 100% FREE BONUS

FOR A LIMITED TIME ONLY – Get the best-selling book *"5 Steps to Learn Absolutely Anything in as Little As 3 Days!"* by Edward Mize absolutely FREE!

Readers who have read this bonus book as well have seen huge increases in their abilities to learn new things and apply it to their lives – so it is *highly recommended* to get this bonus book.

Once again, as a big thank-you for downloading this book, I'd like to offer it to you *100% FREE for a LIMITED TIME ONLY!*

To download your FREE copy, go to:

TeachingNerds.com/Bonus

Chapter 2: Variations of Problems We Go Through

This book works through several different variations of statistical significance problems. Some of the problems are like this one, where you are comparing your measured value with a known baseline value using a known standard deviation.

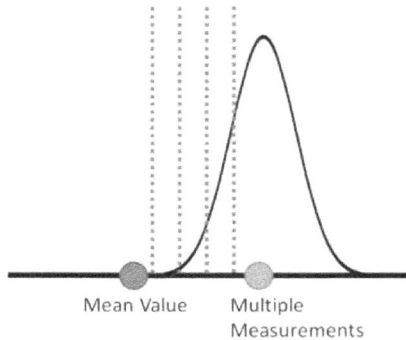

Mean Value   Multiple
              Measurements

However, in other problems, you might not know the baseline value for sure either. The baseline value might have variation in it that might or might not match the variation in your measured values. That would look something like this

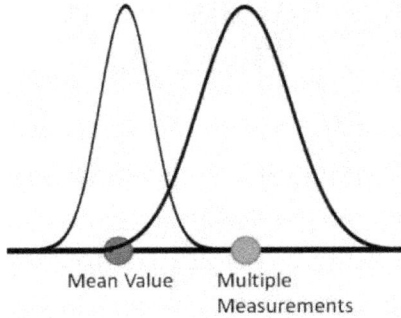

Mean Value   Multiple
               Measurements

In other examples, not only might you have unknown variance in the mean value, you might also have variance in your standard deviation. That would cause the shape of the normal curve to shift and put more probability towards the tails of the curves and would look something like the lower, more spread out orange line below.

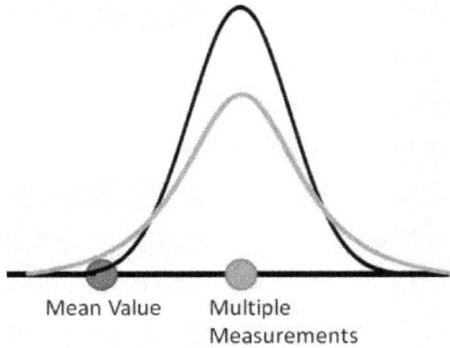

Mean Value   Multiple
               Measurements

Chapter 3: The Process

The overall process of finding statistical significance

1. Determine how many standard deviations our outcome is from the average outcome.

2. Look up that level of standard deviation in a probability table to find how unlikely it is that the outcome would occur by random chance.

Depending on the data in question, there are a couple of different ways to find how many standard deviations an outcome is from the mean, and there are a couple of different probability distributions to choose from.

The following examples go through each of those options to help you determine when each different equation is the right one to use.

The first example below shows calculating the statistical significance of a set of data when you already know what the baseline result is. I.e. if you can look up the typical blood pressure results for adults who are 50 years old, and all you need to calculate is how different the results of your test are, you would use this process, which is known as a Z test and demonstrated below.

Example 1 – Z Test

In this example imagine you are working at a hospital looking at the birth weights of the 30 most recent male babies at your hospital. You would like to conclude that mothers give birth to heavier babies at your hospital than they do nationwide, on average. (Doubtless, this is due to your exceptional care, but since that is hard to prove, let's start by just determining if the babies are heavier).

The average birth weight for an American boy is 7.5 lbs., with a standard deviation of 1.25 lbs.

Using the table below, can you determine, with at least 95% confidence, if male babies born at your hospital exceed the national average weight?

Baby Number	Weight (lbs)
1	5.5
2	5.9
3	6
4	6
5	6.1
6	6.5
7	6.5
8	6.8
9	7
10	7.2
11	7.4
12	7.5
13	7.5
14	7.6
15	7.7
16	7.7
17	7.8
18	7.9
19	8
20	8
21	8.2
22	8.3
23	8.5
24	8.9
25	9
26	9.1
27	9.4
28	10
29	11
30	12
Average	7.8333

The Z Test Equation

The process for finding the solution to this problem is:

1. Determine if a Z test is appropriate to use, or if not, what test should be used

2. Find the Z statistic

3. Determine if we need a 1 tailed or a 2-tailed p-value

4. Find the p-value from the Z table, and compare to our desired confidence level

Step 1 – Determine What Test to Use

As we stated in the introduction, there are a couple variations of statistical significance testing, depending on the parameters of your data set. In this case, we were supplied with the baseline mean and standard deviation data, which means we don't need to calculate those values from our own data set. We also have a moderate number of data points, 30 in this case.

The reason we are using a Z test instead of a T-test like we do in the later examples is that the standard deviation of the population is provided for us, and we expect that our sample matches the population. Some sources will state that you can use a Z-test if you have more than 20-50 data points. (The exact number is a bit of a judgment call). This is not exactly true. What is actually occurring is that as the number of samples increases, the T test converges on the Z test. We will get into that in more detail in later examples. For now, let's see how to apply the Z test to this example.

Step 2 – Find the Test Statistic

The equation for the Z Statistic is

$$Z = \frac{\bar{X} - u_0}{\frac{\sigma}{\sqrt{n}}}$$

Where:

- $\bar{X}$ is the sample mean

- u_0 is the population mean

- σ is the population standard deviation

- n is the test sample size

Note the sigma divided by the square root of n in this equation. This is the standard deviation of the average of the data that we discussed earlier, in the example with the average rolls when rolling multiple dice.

$$Z = \frac{\bar{x} - u_O}{\boxed{\dfrac{\sigma}{\sqrt{n}}}}$$

Std Deviation
Of The Mean

This is the typical form of the Z equation. Personally, however, I prefer not to have the fraction within a fraction. So, if you pull the square root of n to the top of the equation, it looks like this.

$$Z = \frac{\bar{x} - u_o}{\sigma} * \sqrt{n}$$

To me, this makes it clearer that a larger sample size (bigger n) will serve to increase the Z value.

The data from the problem was:

- The population mean value, u_o, is 7.5 lbs.

- The population standard deviation, σ, is 1.25 lbs.

- The number of samples, n, is 30

- The average of the samples, $\bar{x}$, is 7.8333

The resulting equation is

$$Z = \frac{7.8333 - 7.5}{1.25} * \sqrt{30}$$

Which becomes

$$Z = \frac{.3333}{1.25} * 5.478 = 1.461$$

So, the calculated Z statistic is   1.461

Step 3 - 1 Tailed or 2 Tailed

1 tailed or 2 tailed is asking the question if you just want to find out if you are different than the baseline value, or if you care which side of the mean your measured average is on.

For instance, if you want to determine if filling your pickup truck with rocks changes your gas mileage, you would use a 2 tailed distribution.

Theoretically, it could improve or reduce your gas mileage. If you want to determine if the rocks <u>improve</u> your gas mileage, you would use a 1 tailed distribution. (Side note, they don't)

Since the problem asks us to conclude if we are greater than the average, not just different than the average, we will use a 1 tailed Z value. 1 tailed vs 2 tailed is shown graphically below.

Step 4 - Look up our Z value in the Z table, and get the *p*-value

One thing to be cautious of when extracting results from a Z table is exactly what kind of table you are using. What table to use, or how to use it, is partly based on if you need a 1 tailed or 2 tailed result. All Z tables are based on the same standard normal curve. A standard normal curve is a normal curve with a mean of zero and a standard deviation of 1.0

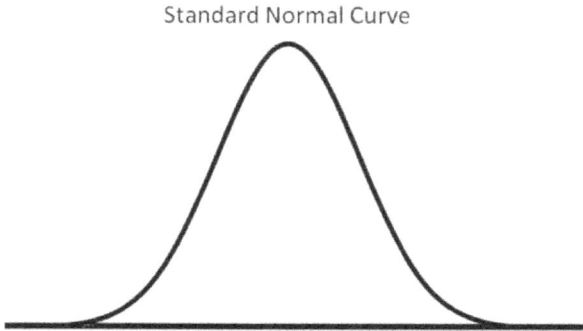

Standard Normal Curve

All Z tables have the same source of data. However, that data can be used in different ways. Some tables can show values filled in from the left

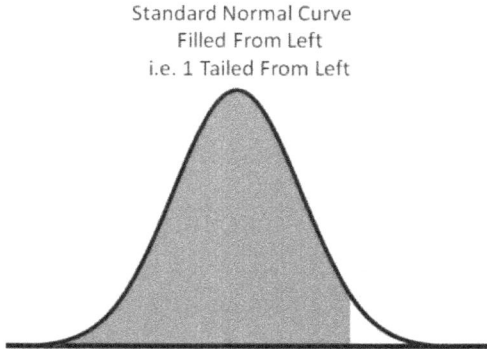

Standard Normal Curve
Filled From Left
i.e. 1 Tailed From Left

Others filled in from the middle

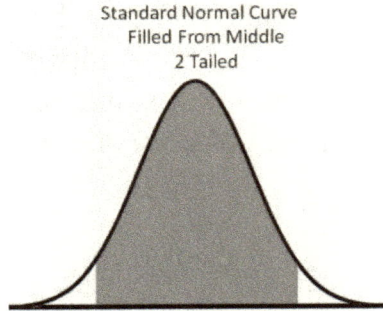

Standard Normal Curve
Filled From Middle
2 Tailed

Some might be filled in from both sides, instead of a single side

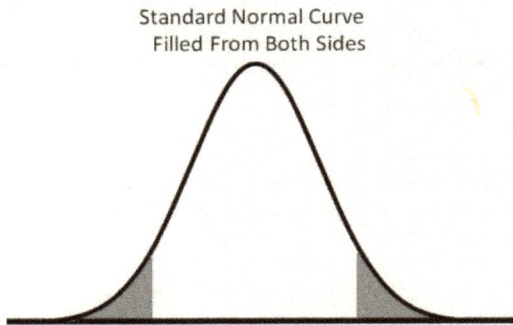

Standard Normal Curve
Filled From Both Sides

Typically, if the table you are looking at doesn't have what you need, you can convert it to a different format by subtracting the confidence values from 1, or by multiplying or dividing them by 2.

In this case, we are trying to determine if we are greater than the average. We want a Z table equivalent to this graph

Standard Normal Curve
Filled From Left

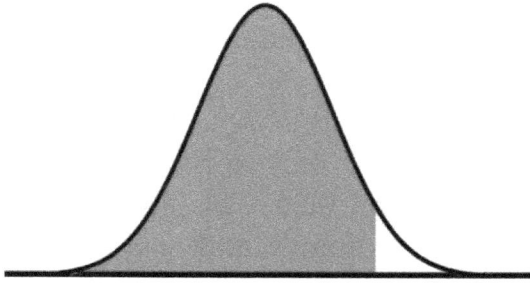

And we want to determine if our data is far enough to the right that 95% of the area under the curve is filled in.

This is a 1 tailed Z Table, from the left

Z	0.00	0.01	0.02	0.03	0.04	0.05	0.06	0.07	0.08	0.09
0.00	0.5000	0.5040	0.5080	0.5120	0.5160	0.5199	0.5239	0.5279	0.5319	0.5359
0.10	0.5398	0.5438	0.5478	0.5517	0.5557	0.5596	0.5636	0.5675	0.5714	0.5753
0.20	0.5793	0.5832	0.5871	0.5910	0.5948	0.5987	0.6026	0.6064	0.6103	0.6141
0.30	0.6179	0.6217	0.6255	0.6293	0.6331	0.6368	0.6406	0.6443	0.6480	0.6517
0.40	0.6554	0.6591	0.6628	0.6664	0.6700	0.6736	0.6772	0.6808	0.6844	0.6879
0.50	0.6915	0.6950	0.6985	0.7019	0.7054	0.7088	0.7123	0.7157	0.7190	0.7224
0.60	0.7257	0.7291	0.7324	0.7357	0.7389	0.7422	0.7454	0.7486	0.7517	0.7549
0.70	0.7580	0.7611	0.7642	0.7673	0.7704	0.7734	0.7764	0.7794	0.7823	0.7852
0.80	0.7881	0.7910	0.7939	0.7967	0.7995	0.8023	0.8051	0.8078	0.8106	0.8133
0.90	0.8159	0.8186	0.8212	0.8238	0.8264	0.8289	0.8315	0.8340	0.8365	0.8389
1.00	0.8413	0.8438	0.8461	0.8485	0.8508	0.8531	0.8554	0.8577	0.8599	0.8621
1.10	0.8643	0.8665	0.8686	0.8708	0.8729	0.8749	0.8770	0.8790	0.8810	0.8830
1.20	0.8849	0.8869	0.8888	0.8907	0.8925	0.8944	0.8962	0.8980	0.8997	0.9015
1.30	0.9032	0.9049	0.9066	0.9082	0.9099	0.9115	0.9131	0.9147	0.9162	0.9177
1.40	0.9192	0.9207	0.9222	0.9236	0.9251	0.9265	0.9279	0.9292	0.9306	0.9319
1.50	0.9332	0.9345	0.9357	0.9370	0.9382	0.9394	0.9406	0.9418	0.9429	0.9441
1.60	0.9452	0.9463	0.9474	0.9484	0.9495	0.9505	0.9515	0.9525	0.9535	0.9545
1.70	0.9554	0.9564	0.9573	0.9582	0.9591	0.9599	0.9608	0.9616	0.9625	0.9633
1.80	0.9641	0.9649	0.9656	0.9664	0.9671	0.9678	0.9686	0.9693	0.9699	0.9706
1.90	0.9713	0.9719	0.9726	0.9732	0.9738	0.9744	0.9750	0.9756	0.9761	0.9767
2.00	0.9772	0.9778	0.9783	0.9788	0.9793	0.9798	0.9803	0.9808	0.9812	0.9817
2.10	0.9821	0.9826	0.9830	0.9834	0.9838	0.9842	0.9846	0.9850	0.9854	0.9857
2.20	0.9861	0.9864	0.9868	0.9871	0.9875	0.9878	0.9881	0.9884	0.9887	0.9890
2.30	0.9893	0.9896	0.9898	0.9901	0.9904	0.9906	0.9909	0.9911	0.9913	0.9916
2.40	0.9918	0.9920	0.9922	0.9925	0.9927	0.9929	0.9931	0.9932	0.9934	0.9936
2.50	0.9938	0.9940	0.9941	0.9943	0.9945	0.9946	0.9948	0.9949	0.9951	0.9952
2.60	0.9953	0.9955	0.9956	0.9957	0.9959	0.9960	0.9961	0.9962	0.9963	0.9964
2.70	0.9965	0.9966	0.9967	0.9968	0.9969	0.9970	0.9971	0.9972	0.9973	0.9974
2.80	0.9974	0.9975	0.9976	0.9977	0.9977	0.9978	0.9979	0.9979	0.9980	0.9981
2.90	0.9981	0.9982	0.9982	0.9983	0.9984	0.9984	0.9985	0.9985	0.9986	0.9986
3.00	0.9987	0.9987	0.9987	0.9988	0.9988	0.9989	0.9989	0.9989	0.9990	0.9990
3.10	0.9990	0.9991	0.9991	0.9991	0.9992	0.9992	0.9992	0.9992	0.9993	0.9993
3.20	0.9993	0.9993	0.9994	0.9994	0.9994	0.9994	0.9994	0.9995	0.9995	0.9995
3.30	0.9995	0.9995	0.9995	0.9996	0.9996	0.9996	0.9996	0.9996	0.9996	0.9997
3.40	0.9997	0.9997	0.9997	0.9997	0.9997	0.9997	0.9997	0.9997	0.9997	0.9998

To read a Z table, you look up the whole number and first decimal of the Z value from the leftmost column and match that with the 2ⁿᵈ decimal of the Z value from the top row.

Since our Z statistic is 1.46, we look up 1.40 going down the left-hand column, and .06 going across the top row. Where that column and row matches is the cumulative probability which is .9279 in this case.

Z Table										
z	0.00	0.01	0.02	0.03	0.04	0.05	0.06	0.07	0.08	0.09
0.00	0.5000	0.5040	0.5080	0.5120	0.5160	0.5199	0.5239	0.5279	0.5319	0.5359
0.10	0.5398	0.5438	0.5478	0.5517	0.5557	0.5596	0.5636	0.5675	0.5714	0.5753
0.20	0.5793	0.5832	0.5871	0.5910	0.5948	0.5987	0.6026	0.6064	0.6103	0.6141
0.30	0.6179	0.6217	0.6255	0.6293	0.6331	0.6368	0.6406	0.6443	0.6480	0.6517
0.40	0.6554	0.6591	0.6628	0.6664	0.6700	0.6736	0.6772	0.6808	0.6844	0.6879
0.50	0.6915	0.6950	0.6985	0.7019	0.7054	0.7088	0.7123	0.7157	0.7190	0.7224
0.60	0.7257	0.7291	0.7324	0.7357	0.7389	0.7422	0.7454	0.7486	0.7517	0.7549
0.70	0.7580	0.7611	0.7642	0.7673	0.7704	0.7734	0.7764	0.7794	0.7823	0.7852
0.80	0.7881	0.7910	0.7939	0.7967	0.7995	0.8023	0.8051	0.8078	0.8106	0.8133
0.90	0.8159	0.8186	0.8212	0.8238	0.8264	0.8289	0.8315	0.8340	0.8365	0.8389
1.00	0.8413	0.8438	0.8461	0.8485	0.8508	0.8531	0.8554	0.8577	0.8599	0.8621
1.10	0.8643	0.8665	0.8686	0.8708	0.8729	0.8749	0.8770	0.8790	0.8810	0.8830
1.20	0.8849	0.8869	0.8888	0.8907	0.8925	0.8944	0.8962	0.8980	0.8997	0.9015
1.30	0.9032	0.9049	0.9066	0.9082	0.9099	0.9115	0.9131	0.9147	0.9162	0.9177
1.40	0.9192	0.9207	0.9222	0.9236	0.9251	0.9265	0.9279	0.9292	0.9306	0.9319
1.50	0.9332	0.9345	0.9357	0.9370	0.9382	0.9394	0.9406	0.9418	0.9429	0.9441

That means 92.79% of the area is at a Z value less than the value we calculated for our data.

Subtracting .9279 from 1 results in a 1 tailed *p*-value of .0721. Since we want 95% confidence or greater, we would need a *p*-value of .05 or less to satisfy that condition. So, we cannot say with 95% confidence that the weights in this hospital are greater than average.

What Can Change the Results

We did not get our desired 95% confidence. How can we get higher confidence? Let's work the problem backward to see what would need to change for us to have 95% confidence that our hospital had heavier babies.

That result would require a value of .95 on the Z table. That value equates to having a Z value of at least 1.65, which we know by looking up .95 and the Z table and seeing the column and row Z values that generate the .95 confidence.

Z	0.00	0.01	0.02	0.03	0.04	0.05	0.06	0.07	0.08	0.09
0.00	0.5000	0.5040	0.5080	0.5120	0.5160	0.5199	0.5239	0.5279	0.5319	0.5359
0.10	0.5398	0.5438	0.5478	0.5517	0.5557	0.5596	0.5636	0.5675	0.5714	0.5753
0.20	0.5793	0.5832	0.5871	0.5910	0.5948	0.5987	0.6026	0.6064	0.6103	0.6141
0.30	0.6179	0.6217	0.6255	0.6293	0.6331	0.6368	0.6406	0.6443	0.6480	0.6517
0.40	0.6554	0.6591	0.6628	0.6664	0.6700	0.6736	0.6772	0.6808	0.6844	0.6879
0.50	0.6915	0.6950	0.6985	0.7019	0.7054	0.7088	0.7123	0.7157	0.7190	0.7224
0.60	0.7257	0.7291	0.7324	0.7357	0.7389	0.7422	0.7454	0.7486	0.7517	0.7549
0.70	0.7580	0.7611	0.7642	0.7673	0.7704	0.7734	0.7764	0.7794	0.7823	0.7852
0.80	0.7881	0.7910	0.7939	0.7967	0.7995	0.8023	0.8051	0.8078	0.8106	0.8133
0.90	0.8159	0.8186	0.8212	0.8238	0.8264	0.8289	0.8315	0.8340	0.8365	0.8389
1.00	0.8413	0.8438	0.8461	0.8485	0.8508	0.8531	0.8554	0.8577	0.8599	0.8621
1.10	0.8643	0.8665	0.8686	0.8708	0.8729	0.8749	0.8770	0.8790	0.8810	0.8830
1.20	0.8849	0.8869	0.8888	0.8907	0.8925	0.8944	0.8962	0.8980	0.8997	0.9015
1.30	0.9032	0.9049	0.9066	0.9082	0.9099	0.9115	0.9131	0.9147	0.9162	0.9177
1.40	0.9192	0.9207	0.9222	0.9236	0.9251	0.9265	0.9279	0.9292	0.9306	0.9319
1.50	0.9332	0.9345	0.9357	0.9370	0.9382	0.9394	0.9406	0.9418	0.9429	0.9441
1.60	0.9452	0.9463	0.9474	0.9484	0.9495	0.9505	0.9515	0.9525	0.9535	0.9545
1.70	0.9554	0.9564	0.9573	0.9582	0.9591	0.9599	0.9608	0.9616	0.9625	0.9633
1.80	0.9641	0.9649	0.9656	0.9664	0.9671	0.9678	0.9686	0.9693	0.9699	0.9706
1.90	0.9713	0.9719	0.9726	0.9732	0.9738	0.9744	0.9750	0.9756	0.9761	0.9767

The equation that we used to calculate the Z value was

$$Z = \frac{\bar{x} - u_o}{\sigma} * \sqrt{n}$$

We see a couple of things. The $\bar{x}$ and square root of n relate to our test results. The u_o and σ values relate to the whole population. We can assume that we cannot affect the population as a whole (for instance by making every other baby born be lighter). So, the numbers we can affect are what is occurring in our hospital.

To be more confident that our hospital has higher weight babies than average, we either need to make our babies even heavier compared to the average, or increase the number of samples. Of those two options, the most viable is probably increasing the number of samples.

If we leave every other number the same, to get a Z value of 1.65 we can solve for the required n value.

$$1.65 = \frac{7.8333 - 7.5}{1.25} * \sqrt{n}$$

This results in an n value of 38.3. Since we can't have a partial sample, this would require 39 samples. We already have 30 samples, so if we measured 9 more babies and got the same average we have now, we could make the claim with 95% confidence that babies born in our hospital were heavier than the national average. (Of course, when making those 9 measurements our average could go up or down.)

The Diminishing Return of Increased Samples

Although in this example we were able to reach our desired confidence level by making additional measurements, it is worth knowing that the effect of increased measurements has a diminishing return. This is due to the square root in front of the n. As you increase n from 1 to 100, this is how the square root of n effect would change

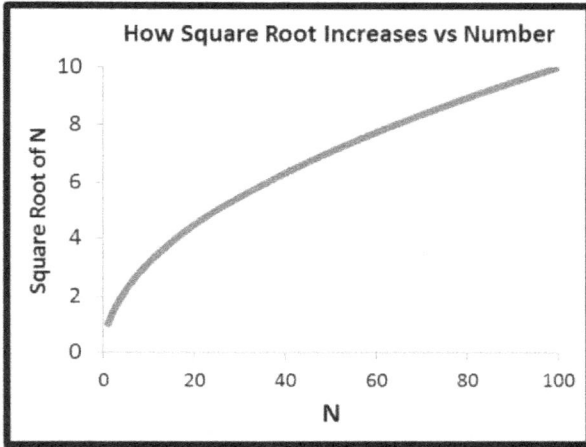

Each subsequent doubling in the value of the square root requires 4 times as many measurements.

Example 1 Summary Table

A table summarizing how we solved this example is shown below. To get the p-value from the Z table in Excel, I used the function

=NORM.S.DIST(xxx,TRUE)

Solution By Hand		
Number of samples	30	
Mean Weight	7.83333	
Population Mean Weight	7.5	provided in problem
Population Standard Deviation	1.25	provided in problem
Difference of Means	0.333	

$$Z = \frac{\bar{x} - u_o}{\sigma} * \sqrt{n}$$

Z Statistic	1.4606	
Cumulative p value	0.9279	
		no, .072 > .05, so you can not make that
1 Tailed p value	0.0721	conclusion

Doing the Z Test in Excel

So far, we saw how to calculate the Z value and turn that into a confidence level. But sometimes you just want an answer. There are some statistical software packages that you could use. But the software you are most likely to have is Excel. If you just want to do the Z test in Excel, it is quite easy. You just need the function

= Z.Test(array, x, [sigma])

Where:

- Array is the data to test, in this case, it would be the 30 sample weights from the hospital

- x is the population mean, in this case, the 7.5 lbs. average population weight

36

- [sigma] is the population standard deviation, in this case, 1.25 lbs. Sigma is in brackets because it is optional. If you don't include it, Excel will use the sample standard deviation from your sample data.

The *p*-value from the Z test in Excel gives the same result as the hand calculation

```
            Excel Solution

    1 Tailed p  value      0.0721
```

Get the Data & Excel Functions

If you want the excel file that contains all of the example data and solutions, it can be downloaded for Free here http://www.fairlynerdy.com/statistical-significance-examples/

Chapter 4: The Normal Curve

What is the physical meaning of the normal curve? And what is the math to generate it?

After all, many constants that we use every day have physical meaning beyond their mere value. Pi is 3.14159…, but it is also the ratio of the circumference of a circle to the diameter of that circle

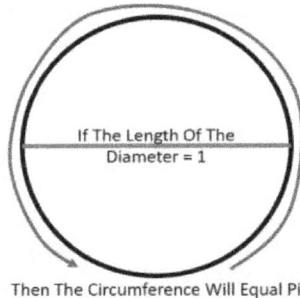

If The Length Of The Diameter = 1

Then The Circumference Will Equal Pi

Since you don't have an infinitely fine ruler to measure the diameter and the circumference, one of the many ways to calculate Pi is
$$Pi = 4 * (1/1 - 1/3 + 1/5 - 1/7 + 1/9 - 1/11 + 1/13 \ldots\ldots)$$

e is another commonly used number. You see the exponential e appear in many places. It has the value of 2.71828…. But it also has a physical meaning. If you have a unit of growth, and then cut the intervals you are compounding on to infinitely small, you get e.

For instance, if you have a 100% interest rate, and compound 1 time, you end up with 200% of your starting money. If you have a 50% interest rate, and it compounds twice, you end up with 225%.

If you keep cutting the time intervals down infinitely fine, eventually you will plateau at a growth of e. This is why e is used to calculate continuously compounded interest. (This site has a great explanation of e https://betterexplained.com/articles/an-intuitive-guide-to-exponential-functions-e/).

You can calculate e by calculating the value of $(1 + 1/n)^n$ as n goes to infinity. If you do that for the first couple values of n you get

n = 1 -> $(1 + 1.0/1)^1 = 2$
n = 2 -> $(1 + 1.0/2)^2 = 2.25$
n = 3 -> $(1 + 1.0/3)^3 = 2.37037$
n = 4 -> $(1 + 1.0/4)^4 = 2.44141$

When n is infinite you get e

But his is a book about statistical significance. So, what about the normal curve? What does it mean, and how can you calculate it?

It turns out that the normal curve was originally developed to approximate the Binomial Theorem. Which tells you how likely an outcome is after a number of discrete trials.

For instance, if I flip a coin 10 times, I can count how many times I get 0 heads, 1 head, 2 heads, etc. If I plot that in a histogram, it is similar to the normal curve. The more discrete events I do, the more it approaches the normal curve. After infinitely many events, it matches the normal curve precisely.

The normal distribution shows up in many more real-world locations than just coin flips. For instance, a plant might have 10 genes that control its height. For any of those 10 genes, it could get the shorter version of the gene or the taller gene. This is a binomial event, and as a result, the heights of the plants will tend to follow the normal distribution. So, if you had 2^{10} plants (1024 plants) on average you would have on average:

- 1 plant with 0 tall genes

- 10 plants with 1 tall gene

- 45 plants with 2 tall genes

- 120 plants with 3 tall genes

- 210 plants with 4 tall genes

- 252 plants with 5 tall genes

- 210 plants with 6 tall genes

- 120 plants with 7 tall genes

- 45 plants with 8 tall genes

- 10 plants with 9 tall genes

- 1 plant with 10 tall gents

A histogram of that distribution is shown below

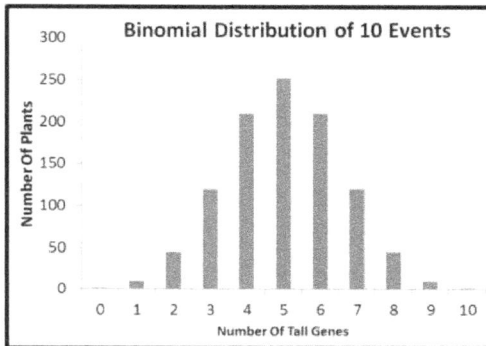

Binomial Distribution of 10 Events

This distribution has an average value of 5 tall genes and a standard deviation of 1.58 tall genes. If we change this into a probability distribution by dividing all the numbers by 1024 total plants, we can compare the result again a normal curve. The normal curve we are comparing against was generated with a mean value of 5 and a standard deviation of 1.5. This is shown below.

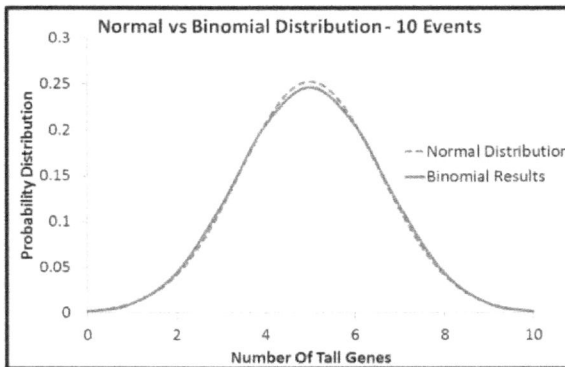

Normal vs Binomial Distribution - 10 Events

If you can't see the two lines in the chart above, it is because they are nearly right on top of each other. And that is after only 10 events. With sufficient additional events, the binomial distribution and the normal distribution become indistinguishable. (For more about the binomial distribution you might check out my book "Probability With The Binomial Distribution And Pascal's Triangle").

So that is an application of the normal curve in real life. It is grounded in the reality of the typical outcomes of events when subjected to random chance. But how can you calculate the normal curve without simulating infinitely many random events? As it turns out, there is an equation for the normal curve, and that is shown below.

$$\varphi(x) = \frac{e^{-\frac{1}{2}x^2}}{\sqrt{2\pi}}$$

This is a probability density function. The factor of square root of 2 Pi at the bottom of that equation ensures that the total area of the normal curve is one. The one-half on the top of the equation ensures that this curve has a standard deviation of 1, and is hence a standard normal curve.

This equation will give a maximum value of one divided by the square root of 2 Pi at a value of x equal to zero.

Doing A T-Test, Which Is Slightly Different Than A Z Test

Let's do another example that has one big difference from the first one. That difference is that the standard deviation of the population is not provided. In real life, this will probably be more common, because really, how many things can you just look up the standard deviation to?

Let's say you are testing if a new fuel additive makes your fleet of trucks more fuel efficient. Will you be able to look up what the standard deviation of the fuel efficiency of those trucks is? No, you'll just have to measure it and calculate the standard deviation based on those results.

This type of test, where you are using a standard deviation that you calculate, is a T-test, as opposed to a Z-test. This test is also knowns as a "Student's T-test", not because it is primarily used by students, but because the developer of the test in the early 1900's developed it under the pseudonym of "Student."

In addition to calculating your own standard deviation, the other big difference between a Z-test and a T-test is that all Z-tests use the standard normal distribution, and the T-test uses a different distribution that varies depending on how much data that you have, a metric known as degrees of freedom (df). The reason that the T-test distribution changes based on how much data you have is to account for the uncertainty you have in calculating your standard deviation.

Let's say that you have 10 data points and you calculate the standard deviation of those 10 points. How do you know if that standard deviation matches the population as a whole? You don't. Your sample standard deviation could be high or it could be low.

The chart below demonstrates the uncertainty that you have in getting standard deviations from measurements. The data was generated by taking the standard deviation of a set of random integers between 1 and 10. The true standard deviation of that set should be the standard deviation of [1, 2, 3, 4, 5, 6, 7, 8, 9, 10] which is 2.872.

However, I don't have the true data set, what I have are measurements drawn from that data set. For one line in the chart below I selected 5 random numbers between 1 and 10 and calculated the standard deviation of those 5 measurements. For the other line, I selected 10 random numbers between 1 and 10 and calculated the standard deviation of those 10 measurements. I did that random selection 1,000 times for both groups and made a histogram of the calculated standard deviations

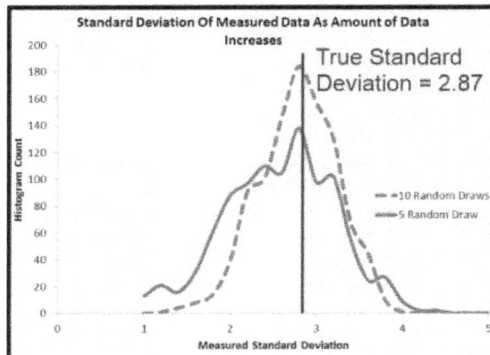

The reason those curves are bumpy instead of smooth arcs is that the chart above was made with data generated from random numbers, as opposed to actual calculated values.

As you can see, in both cases there were times when the data overestimated the population standard deviation and times when it underestimated the standard deviation. However, the examples where there were 10 data points were chosen had a greater probability of being closer to the actual standard deviation.

The T-test accounts for this variance in the measured standard deviation by changing the shape of the T distribution curve depending on how much data you have measured (the degrees of freedom). The less data that you have the more probability is put towards the tails of the curves in the T-distribution compared to the than the standard normal distribution. This is shown in the chart below, where the T-distribution with 2 degrees of freedom (which is quite low) has a lot more outlying probability than the Z-distribution.

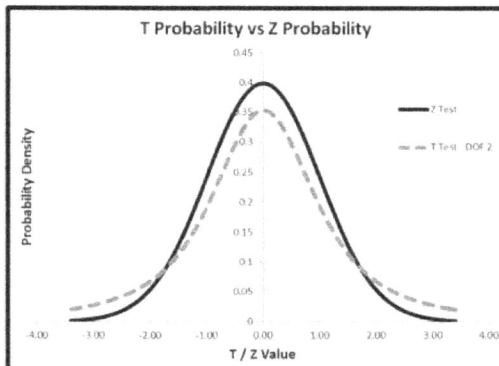

The smaller the degrees of freedom in your test (I.e. derived from a smaller data set) the more heavily the T-distribution weights the tails of the curves. The chart below shows how a T-distribution with a degree of freedom 5 has less probability at the tails of the curves than a degree of freedom 2 distribution.

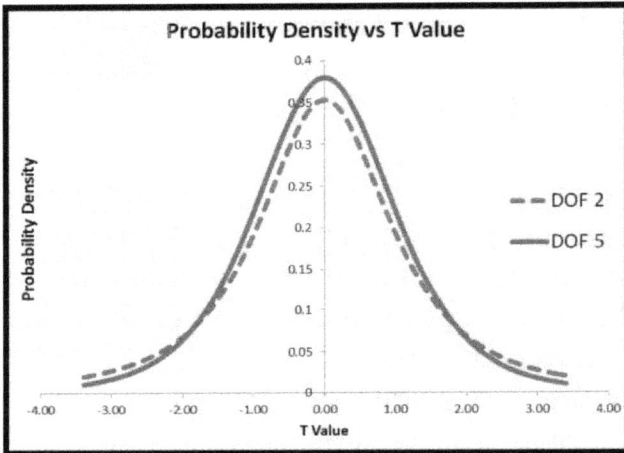

As you increase degrees of freedom, the T-distribution matches the normal distribution more closely. At approximately df = 30, there is very little difference between the two distributions to the eye. With a large enough sample, the T-distribution will eventually exactly match the normal distribution. The chart below shows the T distribution with a degree of freedom 2, 5, 30, and 100.

There is a large change in distribution going from 2 to 5, and from 5 to 30; however, there is only a little change in the probability distribution going from 30 to 100. This is because the T-distribution with 30 degrees of freedom already closely matches the standard normal distribution

(Z distribution), so additional data samples that increase the degrees of freedom only make small changes.

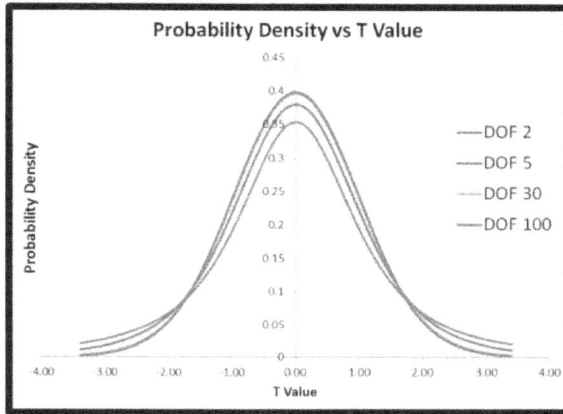

With a large enough sample, essentially what you have done is recreate the population standard deviation.

Summary of a T-test

- In a T-test, you have to use the sample data to determine the standard deviation of the population, as opposed to it being provided

- Because of the uncertainty in the standard deviation, the T-distribution puts more probability at the tails of the probability distribution

- With enough data, the T-distribution exactly matches the standard normal distribution. (i.e. you could use a T-table or a Z-table and get the same result)

- Enough data is usually assumed to be a df somewhere between 20 and 50. (That value is something of a judgment call. If you want to see how T-distribution changes as you change the degrees of freedom, this online tool is useful. http://rpsychologist.com/d3/tdist/)

The following example shows the use of the 1 sample T-test.

Example 2 – 1 Sample T-Test

You have been told that the average height of female college students in the United States is 5.5 feet. You have measured 15 students at your college, and want to determine if the average height at your college is statistically different than the average across the U.S.

Student Number	Height (feet)
1	5.1
2	5.2
3	5.4
4	5.5
5	5.5
6	5.6
7	5.6
8	5.7
9	5.7
10	5.8
11	5.8
12	5.9
13	6
14	6.1
15	6.2

Here we know the population mean, 5.5 feet, but not the population standard deviation. We will have to derive that from our data. (With this type of test there is an assumption that the population standard deviation is not systematically different from the data that we have. For instance, we didn't intentionally admit students based on their height in order to make a narrow distribution.)

The process for finding the solution to this problem is very similar to what we did for the Z test.

1. Determine what test to use

2. Find the test statistic

3. Determine if we need a 1 tailed or a 2-tailed p-value

4. Find the p-value from the table, and compare to our desired confidence level.

Solving The 1 Sample T-Test

Step 1 – Determine What Test to Use

We have 1 sample set of data. That sample set is only 15 measurements. We have a population mean, but not a population standard deviation. Since we don't have a population standard deviation, and we have fewer than 20 samples, we will want to do a T test. Since we have 1 sample, and we have a population mean, it will be a 1 Sample T Test

These are the key points of our problem

49

- We have exactly 1 set of data. We don't have a before and after. We don't have a test group and a control group. We have 1 and only 1 set of data.

- We do have a population mean to calculate our difference off of

- We do not have a population standard deviation

These factors mean we should do a 1 sample T test.

Keeping track of all the T-test equations, and when you should use them, can get confusing. If you want to get a one page PDF cheat sheet of all the T-test and Z-test equations, and when you would use them, you can get it here http://www.fairlynerdy.com/hypothesis-testing-cheat-sheets/

Step 2 – Find the Test Statistic

The equation for the test statistic is

$$t = \frac{\bar{x} - u_0}{\frac{s}{\sqrt{n}}}$$

Where:

- $\bar{x}$ is the sample mean

- u_0 is the population mean

- s is the sample standard deviation

- n is the test sample size

This could also be re-written with the square root of n pulled out of the denominator and be

$$t = \frac{\bar{x} - u_o}{s} * \sqrt{n}$$

Really this equation isn't different from the Z equation, except we use a sample standard deviation that we calculate instead of the population standard deviation that is provided. In both cases, we are calculating the number of standard deviations outside the population mean that your data is, and factoring that number by the size of the data (square root of n)

To calculate the sample standard deviation in Excel, you use the function =STDEVA(). To do it by hand, you use this equation

$$s = \sqrt{\frac{\sum_{i=1}^{n}(x_i - \bar{x})^2}{n - 1}}$$

What you are doing is:

- Take the difference of each item from the mean

- Square that value

- Sum all of the squares

- Divide by the number of samples minus 1

- Take the square root

Note that what we are using here is the sample standard deviation, which has (n-1) on the denominator, as opposed to population standard deviation which would have n instead of (n-1)

Doing that in a few columns looks like this

Student Number	Height (feet)		Difference From Mean	Squared Difference From Mean
1	5.1		-0.57	0.3287
2	5.2		-0.47	0.2240
3	5.4		-0.27	0.0747
4	5.5		-0.17	0.0300
5	5.5		-0.17	0.0300
6	5.6		-0.07	0.0054
7	5.6		-0.07	0.0054
8	5.7		0.03	0.0007
9	5.7		0.03	0.0007
10	5.8		0.13	0.0160
11	5.8		0.13	0.0160
12	5.9		0.23	0.0514
13	6		0.33	0.1067
14	6.1		0.43	0.1820
15	6.2		0.53	0.2774
Average	5.673			
			Sum Squared Difference From Mean	1.3493
			Sample Standard Deviation	0.2999

So, in this example:

$\bar{x}$ = sample mean= 5.673
u_o = population mean= 5.5 (from problem statement)
s = sample standard deviation= .3105
n = test sample size= 15

The equation for the test statistic is

$$t = \frac{\bar{x} - u_o}{s} * \sqrt{n}$$

When we plug our values in, we get

$$t = \frac{5.673 - 5.5}{.3105} * \sqrt{15} = 2.16$$

So, the T value is 2.16

Since this is a T-test, not a Z test, we cannot use the Z table. Instead, we need to use the T table. A T table is different than the Z table because it puts more probability on the tails of the curve for low sample sizes. To use a T table, we need to calculate degrees of freedom, which is a measure of how large our sample size is.

The equation for degrees of freedom for a 1 sample T test is

$$df = n - 1$$

So, in this case, df = 14

Step 3 - 1 Tailed or 2 Tailed

Once again, we need to decide if we are just looking for any change in results, or if we need change in a certain direction.

Since the problem just asks us if the results are statistically different, we will use a 2 tailed T test. Since the problem didn't specify how confident we needed to be, we will assume that we need at least 95% confidence.

Step 4 – Using A T-Table

With that information, we can reference the T Table

One Tail	0.005	0.01	0.025	0.05	0.1	0.25
Two Tail	0.01	0.02	0.05	0.1	0.2	0.5
Degrees of Freedom						
1	63.6567	31.8205	12.7062	6.3138	3.0777	1.0000
2	9.9248	6.9646	4.3027	2.9200	1.8856	0.8165
3	5.8409	4.5407	3.1824	2.3534	1.6377	0.7649
4	4.6041	3.7469	2.7764	2.1318	1.5332	0.7407
5	4.0321	3.3649	2.5706	2.0150	1.4759	0.7267
6	3.7074	3.1427	2.4469	1.9432	1.4398	0.7176
7	3.4995	2.9980	2.3646	1.8946	1.4149	0.7111
8	3.3554	2.8965	2.3060	1.8595	1.3968	0.7064
9	3.2498	2.8214	2.2622	1.8331	1.3830	0.7027
10	3.1693	2.7638	2.2281	1.8125	1.3722	0.6998
11	3.1058	2.7181	2.2010	1.7959	1.3634	0.6974
12	3.0545	2.6810	2.1788	1.7823	1.3562	0.6955
13	3.0123	2.6503	2.1604	1.7709	1.3502	0.6938
14	2.9768	2.6245	2.1448	1.7613	1.3450	0.6924
15	2.9467	2.6025	2.1314	1.7531	1.3406	0.6912
16	2.9208	2.5835	2.1199	1.7459	1.3368	0.6901
17	2.8982	2.5669	2.1098	1.7396	1.3334	0.6892
18	2.8784	2.5524	2.1009	1.7341	1.3304	0.6884
19	2.8609	2.5395	2.0930	1.7291	1.3277	0.6876
20	2.8453	2.5280	2.0860	1.7247	1.3253	0.6870
25	2.7874	2.4851	2.0595	1.7081	1.3163	0.6844
30	2.7500	2.4573	2.0423	1.6973	1.3104	0.6828

The T table is not the same as a Z Table. A Z table has the Z value along the outside of the table. The T table has degrees of freedom and the *p*-value along the outside of the table. The reason the tables are different is that the T-table is trying to cram an extra piece of information, the degrees of freedom, into the same two-dimensional table. The degree of freedom controls the shape of the probability distribution, so it is required information. That problem of extra information could have been solved by having many different T-tables, on for each df value, but instead, the typical T-table simply shows less dense information than a Z-table.

For instance, on a typical Z-table you can look up a Z-value to a .01 precision

.01 Precision On A Z Table

Z	0.00	0.01	0.02	0.03	0.04	0.05
0.00	0.5000	0.5040	0.5080	0.5120	0.5160	0.5199
0.10	0.5398	0.5438	0.5478	0.5517	0.5557	0.5596
0.20	0.5793	0.5832	0.5871	0.5910	0.5948	0.5987
0.30	0.6179	0.6217	0.6255	0.6293	0.6331	0.6368
0.40	0.6554	0.6591	0.6628	0.6664	0.6700	0.6736

But on this T-table the distribution of the test statistic is much coarser. For instance, on this T-table with a degree of freedom of 14, you cannot look up a T statistic between 2.62 and 2.14

On This T-Table We Only Have Resolution To ~.4

11	3.1058	2.7181	2.2010	1.7959	1.3634	0.6974
12	3.0545	2.6810	2.1788	1.7823	1.3562	0.6955
13	3.0123	2.6503	2.1604	1.7709	1.3502	0.6938
14	2.9768	2.6245	2.1448	1.7613	1.3450	0.6924
15	2.9467	2.6025	2.1314	1.7531	1.3406	0.6912

If we desire a finer resolution than is in the T-table, we could interpolate the values. Otherwise, we can use a source of the T-value other than this table, such as Excel, to get a more precise value.

With 14 degrees of freedom and a T value of 2.16, we look that up in the T table

T Table

.05 p value

Degrees of Freedom	0.005 (One Tail) / 0.01 (Two Tail)	0.01 / 0.02	0.025 / 0.05	0.05 / 0.1	0.1 / 0.2	0.25 / 0.5
1	63.6567	31.8205	12.7062	6.3138	3.0777	1.0000
2	9.9248	6.9646	4.3027	2.9200	1.8856	0.8165
3	5.8409	4.5407	3.1824	2.3534	1.6377	0.7649
4	4.6041	3.7469	2.7764	2.1318	1.5332	0.7407
5	4.0321	3.3649	2.5706	2.0150	1.4759	0.7267
6	3.7074	3.1427	2.4469	1.9432	1.4398	0.7176
7	3.4995	2.9980	2.3646	1.8946	1.4149	0.7111
8	3.3554	2.8965	2.3060	1.8595	1.3968	0.7064
9	3.2498	2.8214	2.2622	1.8331	1.3830	0.7027
10	3.1693	2.7638	2.2281	1.8125	1.3722	0.6998
11	3.1058	2.7181	2.2010	1.7959	1.3634	0.6974
12	3.0545	2.6810	2.1788	1.7823	1.3562	0.6955
13	3.0123	2.6503	2.1604	1.7709	1.3502	0.6938
14	2.9768	2.6245	2.1448	1.7613	1.3450	0.6924
15	2.9467	2.6025	2.1314	1.7531	1.3406	0.6912

14 df → 14

The value of 2.16 isn't in the T table, but the value of 2.144 is close enough. We can read up to see that the two-tailed confidence at this T value is .05. This means that we are at least 95% confident that this college has a statistically different height than the average.

Putting that all into one table

Solution By Hand	
Number of samples	15
Mean Height	5.67333
Sample Standard Deviation	0.3105

$$t = \frac{\bar{x} - u_o}{s} * \sqrt{n}$$

T Statistic	2.16238

$$df = n - 1$$

Degrees of Freedom	14
Cumulative p value	0.0242
2 Tailed p value	0.0484

Yes, we can say with greater than 95% confidence that this college is statistically different

Excel generated the p value using the function

= T.Dist.2T(x, DF)

It returned .0484 which is slightly more precise than the .05 we got from the T-table since we had to go with the closest value of 2.14, which didn't precisely match our T value

Doing a 1 sample T Test in Excel

Unfortunately, there is no easy one line way to replicate a 1 sample T test in excel. You are forced to go through it step by step. But using the equation =STDEVA to get the sample standard deviation, and =T.Dist(), T.Dist.RT() or =T.Dist.2T() to lookup a value from the T table makes it relatively painless

2 Tailed vs 1 Tailed Result

Noting that the previous problem asked us if the results were statistically different than the mean, rather than statistically greater was important. It meant that we are asking the question to see if our mean value falls in either of these parts of the probability distribution

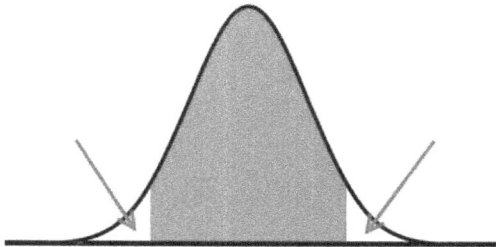

As opposed to asking if it falls in this part

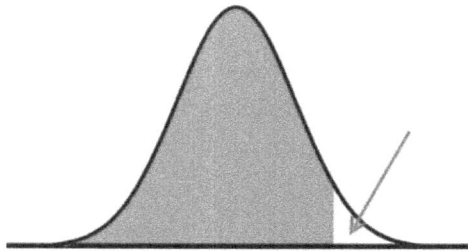

For data that falls in either of these two spots you can get a different result, in terms of do you meet your required confidence level.

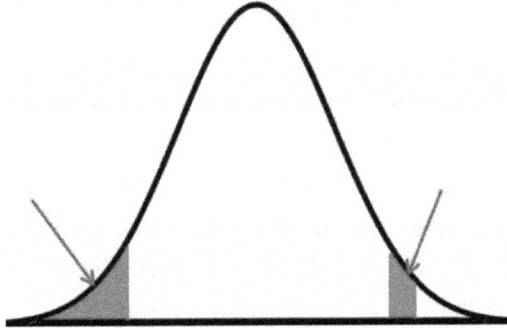

You will always be less confident with a 2-tailed test than a 1 tailed test. (Assuming the 1 tailed test was pointed in the direction of your mean.) The reason is simple if you think about the confidence as area under the curve. With a 2 tail test, you are missing the area on the opposite tail, for any given T or Z value.

How Much Data Do You Need?

One question you might ask when working with data is "How much data do I need to get a statistically significant result?"

With a Z test our answer was "At least 1 data point, depending on how different those results are from the baseline results" With a T-test our answer is "At least 2 data points, depending…."

Why do you need at least 2 data points for a T-test when you only need at least 1 for a Z test?

The answer is found in the degree of freedom and sample standard deviation equations. Both of those contain an "n-1" in the equation.

Degree of freedom equation

$$df = \boxed{n - 1}$$

Standard Deviation Equation

$$s = \sqrt{\frac{\sum_{i=1}^{n}(x_i - \bar{x})^2}{\boxed{n - 1}}}$$

With only 1 data point that would result in a zero. A zero degree of freedom results in an undefined T-distribution. It also means a zero is on the denominator of the standard deviation equation, which does not work.

More intuitively, standard deviation measures how far data points are spread from the mean. With one point, there can be no spread and the standard deviation is undefined. The reason we could get an answer with only 1 data point in the Z test is that our standard deviation came from outside the data. It was provided as part of the problem statement.

The full answer to "How much data do we need to be statistically significant?" depends on how small of a difference we are trying to measure. To conclude that a value is significantly different from the mean requires a lot less data if that difference is large than if it is very small.

Paired T Test – When You Use the Same Test Subject Multiple Times

The big difference between the second example and the first example was that the problem didn't provide a standard deviation in the second example; it needed to be derived from the data. The difference between this third example and the second example is that a mean value isn't provided in this example.

Instead you get two sets of data representing a before and an after. Instead of calculating the difference from the mean, we will calculate how much the "after" data set changed from the "before" data set and determine if that level of change is significant.

The times when you would use this type of test is when you are measuring the exact same individual multiple times (typically twice). For instance, you would use this test if:

- You have a drug trial and measure a person's blood pressure before and after giving them the drug

- You are making a new golf club and measure how far people can hit a golf ball with a standard club vs your new club

- You are a high-priced restaurant consultant, and you measure weekly sales in a restaurant before your visit, and after you come in to give advice and make improvements

You would not use this analysis, and instead use a different equation if:

- You have a drug trial and give one group of people a placebo and the other group your new drug, and measure the blood pressure of both groups.

- You give 10 people your new golf clubs and 10 people your old ones and see how well each group scores on a round of golf

- You measure the sale of restaurants you have consulted, and compare them with similar restaurants in the area that haven't utilized your services yet

The key thing to remember is that this hypothesis test determines if distinct individuals change in the period of time spanning two measurements.

Example 3 – Paired T Test

You weigh 20 people before and after a diet. You want to determine with 99% confidence if the participants lost weight while on the diet.

The key things to note are:

- This is a paired T test because we have a before and after of each individual

- We are testing to see if they lost weight, not just if their weight changed data

Participant #	Before Weight (lbs)	After Weight (lbs)
1	123	116
2	135	131
3	110	102
4	115	108
5	150	140
6	170	172
7	108	113
8	155	144
9	132	139
10	118	109
11	137	125
12	145	140
13	181	175
14	115	115
15	182	173
16	145	137
17	132	127
18	119	116
19	127	125
20	160	145

Solving the Paired T-Test

Step 1 – Determine What Test to Use

For this problem, we will use a Paired T test. The reason – we have a before measurement, and an after measurement for the same participants.

Step 2 – Find the Test Statistic

The equation for t for a paired T test is

$$t = \frac{\bar{d}}{\sqrt{\dfrac{s^2}{n}}}$$

Where

- $\bar{d}$ is the average of the difference between the after sample and the before sample

- s is the sample standard deviation

- n is the test sample size

This equation could be rewritten to be

$$t = \frac{\bar{d}}{s} * \sqrt{n}$$

Notice that this equation is almost the same as our previous T equation which was

$$t = \frac{\bar{x} - u_o}{s} * \sqrt{n}$$

The only difference is instead of measuring the difference of a data set against a known mean value, we are measuring how much the value changed against another data set. Other than replacing $(\bar{x} - u_o)$ with $\bar{d}$, the equation is exactly the same. And frankly the $\bar{d}$ means almost exactly the same thing as the $(\bar{x} - u_o)$, especially since $\bar{d}$ could be written as $(\bar{x}_{new} - \bar{x}_{old})$

$\bar{d}$ is the average change in weight. Calculating it is as simple as subtracting the old weight from the new weight for each item, and finding the average of that value.

Participant #	Before Weight (lbs)	After Weight (lbs)	Change in Weight
1	123	116	-7
2	135	131	-4
3	110	102	-8
4	115	108	-7
5	150	140	-10
6	170	172	2
7	108	113	5
8	155	144	-11
9	132	139	7
10	118	109	-9
11	137	125	-12
12	145	140	-5
13	181	175	-6
14	115	115	0
15	182	173	-9
16	145	137	-8
17	132	127	-5
18	119	116	-3
19	127	125	-2
20	160	145	-15
Average	137.95	132.6	-5.35

The average of the change in weight for this data is -5.35 lbs.

Sample Standard Deviation

To calculate the sample standard deviation the question is, which set of data should you take the standard deviation on? The before data? Or the after data? The answer is neither.

Take the standard deviation on the difference, the $\bar{d}$ column. Note that this is the real reason that this equation is different from the previous ones. This is the reason you need to subtract each pair of data individually instead of just taking the difference of the mean values.

The sample standard deviation of the change in weight column (found using STDEVA() in excel, but see example 2 for a walkthrough of sample standard deviation) is 5.6127 lbs. Notice that this number is positive even though the average change in weight is -5.35 lbs. Standard deviation is always positive.

Number of Samples

The number of samples is 20. This is a little bit tricky because we actually have 40 measurements. But for this Paired T Test, n is representing each pair of measurements, and we have 20 pairs of measurements.

So, for this example the values are

- $\bar{d} = -5.35$

- $s = 5.6127$

- $n = 20$

$$t = \frac{\bar{d}}{s} * \sqrt{n}$$

Plugging our values into the equation results in

$$t = \frac{-5.35}{5.6127} * \sqrt{20} = -4.2628$$

Since this is a T test, we also need degrees of freedom before we can look up the result in the T table. The equation for degrees of freedom is

$$df = n - 1$$

So, since we have 20 pairs of measurements, the degrees of freedom is 19

Step 3 - 1 Tailed or 2 Tailed

Since we want to determine if the participants lost weight on the diet, and not just if their weight changed, we want a 1 tailed p value. I.e. if they gained a statistically significant amount of weight on this diet (the all chocolate diet) we would get a negative result.

Step 4 - Find the p value from the table, and compare to our desired confidence level

The T value that we have is -4.2628. We are trying to determine with 99% confidence that the new results are less than the baseline results. This is equivalent to determining if at least 99% of the area is to the right of -4.2628 on this T-distribution chart.

T Table

	One Tail	0.005	0.01	0.025	0.05	0.1	0.25
	Two Tail	0.01	0.02	0.05	0.1	0.2	0.5
Degrees of Freedom							
1		63.6567	31.8205	12.7062	6.3138	3.0777	1.0000
2		9.9248	6.9646	4.3027	2.9200	1.8856	0.8165
3		5.8409	4.5407	3.1824	2.3534	1.6377	0.7649
4		4.6041	3.7469	2.7764	2.1318	1.5332	0.7407
5		4.0321	3.3649	2.5706	2.0150	1.4759	0.7267
6		3.7074	3.1427	2.4469	1.9432	1.4398	0.7176
7		3.4995	2.9980	2.3646	1.8946	1.4149	0.7111
8		3.3554	2.8965	2.3060	1.8595	1.3968	0.7064
9		3.2498	2.8214	2.2622	1.8331	1.3830	0.7027
10		3.1693	2.7638	2.2281	1.8125	1.3722	0.6998
11		3.1058	2.7181	2.2010	1.7959	1.3634	0.6974
12		3.0545	2.6810	2.1788	1.7823	1.3562	0.6955
13		3.0123	2.6503	2.1604	1.7709	1.3502	0.6938
14		2.9768	2.6245	2.1448	1.7613	1.3450	0.6924
15		2.9467	2.6025	2.1314	1.7531	1.3406	0.6912
16		2.9208	2.5835	2.1199	1.7459	1.3368	0.6901
17		2.8982	2.5669	2.1098	1.7396	1.3334	0.6892
18		2.8784	2.5524	2.1009	1.7341	1.3304	0.6884
DF 19 → 19		2.8609	2.5395	2.0930	1.7291	1.3277	0.6876
20		2.8453	2.5280	2.0860	1.7247	1.3253	0.6870

However, T and Z tables aren't always set up how you need them. For instance, this T Table doesn't have any negative values in it. That is because it is a table from the left, not the right. It is more common to see them set up from the left, which is equivalent to finding this area

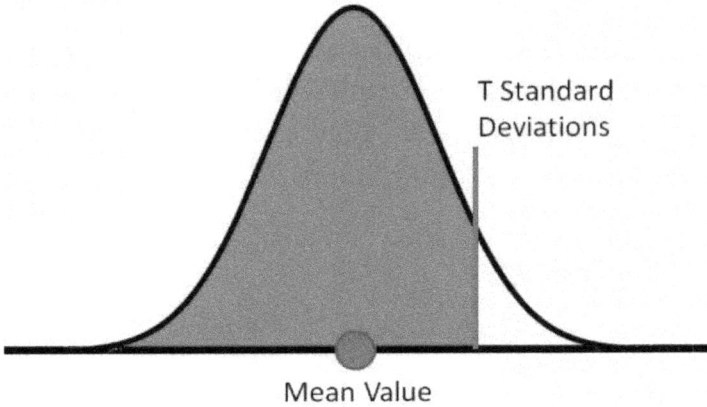

T Standard
Deviations

Mean Value

As opposed to finding this area, if the chart was from the right.

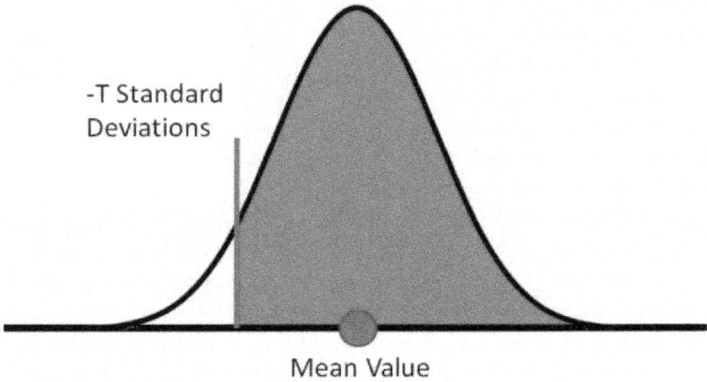

-T Standard
Deviations

Mean Value

However, since the T and Z distributions are symmetric, we can simply change the sign of the T or Z value and use that value. I.e. if you get a value for 2 standard deviations using a chart from the left, it would be the same as the value you would get for -2 standard deviations using a chart from the right.

The other thing we could do with a curve starting at zero from the left (which is not the chart that is shown in this book) would be to find the area to the left of -4.2628, and subtract it from 1.0 to get the area on the right.

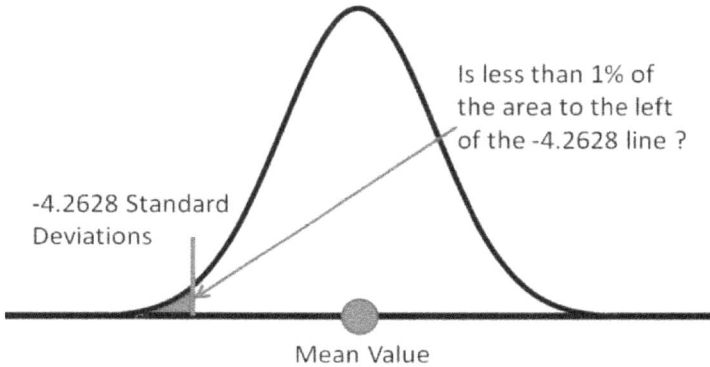

Is less than 1% of the area to the left of the -4.2628 line ?

-4.2628 Standard Deviations

Mean Value

Getting the Confidence Level from Excel

Looking up a T value of 4.2628 and DF of 19 in excel using = T.Dist.RT() we get a cumulative p-value of .0002. In Excel, if I had used -4.2628 I would have gotten the left slice shown in the chart above and would have had to subtract the result from 1.0. Since .0002 is less than the .01 required for 99% confidence, we can say that we are 99% confident that the participants lost weight

I got the .0002 value from Excel. If I had looked it up in a T table, I would have seen that a T value of 4.2628 is greater than 2.8609, which is the T value associated with a 99.5% confidence, so I would have been able to answer the question, but would not have been able to determine that the exact probability was .0002 since the table does not have that level of detail.

Degrees of Freedom	One Tail	0.005	0.01	0.025	0.05	0.1	0.25
	Two Tail	0.01	0.02	0.05	0.1	0.2	0.5

1 tail confidence of 1-.005 = 99.5%

Degrees of Freedom	0.005	0.01	0.025	0.05	0.1	0.25
1	63.6567	31.8205	12.7062	6.3138	3.0777	1.0000
2	9.9248	6.9646	4.3027	2.9200	1.8856	0.8165
3	5.8409	4.5407	3.1824	2.3534	1.6377	0.7649
4	4.6041	3.7469	2.7764	2.1318	1.5332	0.7407
5	4.0321	3.3649	2.5706	2.0150	1.4759	0.7267
6	3.7074	3.1427	2.4469	1.9432	1.4398	0.7176
7	3.4995	2.9980	2.3646	1.8946	1.4149	0.7111
8	3.3554	2.8965	2.3060	1.8595	1.3968	0.7064
9	3.2498	2.8214	2.2622	1.8331	1.3830	0.7027
10	3.1693	2.7638	2.2281	1.8125	1.3722	0.6998
11	3.1058	2.7181	2.2010	1.7959	1.3634	0.6974
12	3.0545	2.6810	2.1788	1.7823	1.3562	0.6955
13	3.0123	2.6503	2.1604	1.7709	1.3502	0.6938
14	2.9768	2.6245	2.1448	1.7613	1.3450	0.6924
15	2.9467	2.6025	2.1314	1.7531	1.3406	0.6912
16	2.9208	2.5835	2.1199	1.7459	1.3368	0.6901
17	2.8982	2.5669	2.1098	1.7396	1.3334	0.6892
18	2.8784	2.5524	2.1009	1.7341	1.3304	0.6884
19	2.8609	2.5395	2.0930	1.7291	1.3277	0.6876
20	2.8453	2.5280	2.0860	1.7247	1.3253	0.6870

df 19→19

4.2628 > 2.8609

A *p*-value of less than .005 would be the best you can extract from this T table due to its coarseness. Excel was able to determine the *p*-value was .0002

Putting it all together into one table

Number of samples	20
Mean Change in Weight	-5.35
Sample Standard Deviation Change in Weight	5.6127

$$t = \frac{\bar{d}}{s} * \sqrt{n}$$

T Statistic	-4.263

$$df = n - 1$$

Degrees of Freedom	19

		Yes, we can say with greater than 99% confidence that
Cumulative p value	0.0002	the participants lost weight
2 Tailed p value	0.0004	

Paired T-Test Excel Solution

You can do a Paired T test in a single line in Excel with the function

=TTest(array1, array2, tails, type)

In this case

array1 = before measurement
array2 = after measurement
tails = 1 tailed or 2 tailed (1 tailed for this problem)
type = what type of T-test (1 for Paired T-Test for this problem)

When you plug in the before and after data into that excel function you get .0002, just like the by hand solution

```
Excel Solution

Paired T Test     0.0002
```

Chapter 5: Is There a Change?

Up until now all the examples we have done have merely asked the question "Is there a statistically significant change?" Sometimes we were interested in the direction of change, other times only if a change existed at all. However, there are times when you are also interested in the magnitude of change.

Imagine that you are the chief data officer for a large chain of stores such as Walmart. You have a new layout for your stores that you think will increase sales. To test the concept, you direct 15 sample stores to change their layout and get before and after results for those stores. Those results show that the new layout increased sales.

However, you need to know something else in addition to the mere direction of change. You need to know if the change increased sales by at least 1 million dollars annually per store. After all, making the switch to the new layout isn't free. It will cost money to change all the layouts, and it is only worth doing of the new revenue is higher than the old revenue by a quantity that is driven by the cost of the project, not just a quantity-driven by statistical significance.

The next example reworks the previous problem to show how you can incorporate a required level of change into the statistical significance calculation.

Example 3A – Paired T-Test with Non-Zero Hypothesis

You weigh 20 people before and after a diet. You want to determine with 95% confidence if the average weight loss was at least 4 pounds. Note, this uses the same data as Example 3, the only difference is instead of trying to determine if the participants lost more than **zero pounds**, we are trying to determine if they lost more than **four pounds**. That subtle change will affect our hand calculation and the one line Excel function.

Data

Participant #	Before Weight (lbs)	After Weight (lbs)
1	123	116
2	135	131
3	110	102
4	115	108
5	150	140
6	170	172
7	108	113
8	155	144
9	132	139
10	118	109
11	137	125
12	145	140
13	181	175
14	115	115
15	182	173
16	145	137
17	132	127
18	119	116
19	127	125
20	160	145

Solving the Paired T Test with A Non-Zero Hypothesis Problem:

Step 1 – Determine What Test to Use

For this problem, we will use a Paired T-test using the same logic as before.

Step 2 – Find the Test Statistic

The equation that we will use for this Paired T-Test with a hypothesized difference is

$$t = \frac{\bar{d} - u_O}{s} * \sqrt{n}$$

Where:

- $\bar{d}$ is the average of the difference between the before sample and after sample

- s is the sample standard deviation

- n is the test sample size

- u_o is the hypothesized mean difference

This is the same equation as before, except that we are subtracting our hypothesized value from the average difference, which will, in this case, reduce the t statistic. (It could theoretically increase the t statistic if the problem were different and was something like "Do you have 95% confidence that the participants either lost weight or gained less than 7 pounds)

Previously the problem was this

Does 99% Of The Area Fall To The Right Of The Old Average?

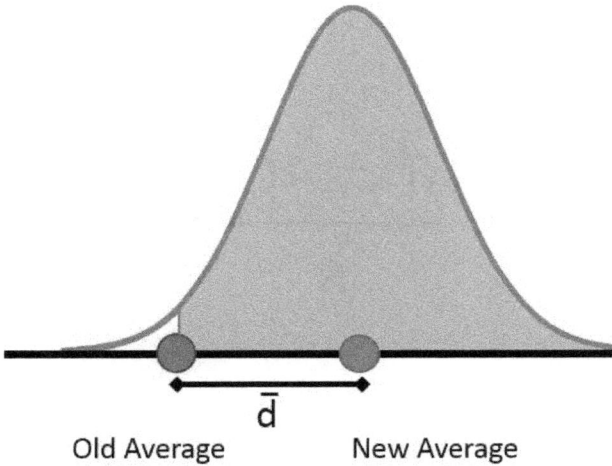

$\overline{d}$

Old Average   New Average

The new problem is this

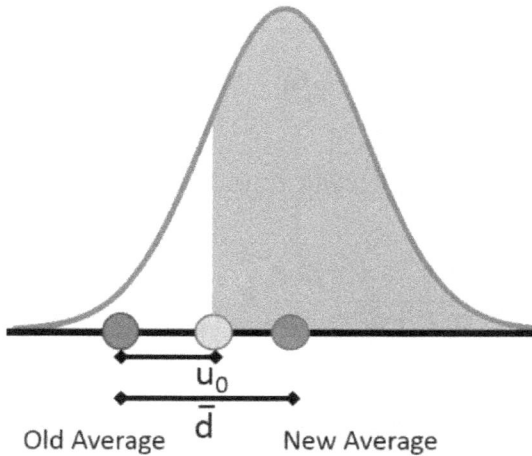

Does 95% Of The Area Fall To The
Right Of The Old Average Plus An Offset?

u_0

$\bar{d}$

Old Average   New Average

The first few steps are the same in example 3. We find the average change in weight to be -5.35 pounds, the standard deviation of the change in weight to be 5.6127 pounds, and we have a sample size of 20.

The only new information is that the hypothesis that we are testing is that the dieters lost at least 4 pounds, which makes u_0 equal to -4

So, the values to use in our equation are:

- $\bar{d} = -5.35$

- $s = 5.6127$

- $n = 20$

- $u_0 = -4$

$$t = \frac{\bar{d} - u_O}{s} * \sqrt{n}$$

Plugging our values into the equation results in

$$t = \frac{-5.35 - (-4)}{5.6127} * \sqrt{20} = -1.0757$$

As you can see, this t statistic of -1.0757 is a lot smaller in magnitude than the previous t statistic of -4.2628. Changing the problem to losing at least 4 pounds has made it a lot more difficult to accomplish.

The equation for degrees of freedom for this problem is the same as Example 3, which is

$$df = n - 1$$

Since we have 20 pairs of measurements, the degrees of freedom is 19

Step 3 - 1 Tailed or 2 Tailed

Since we want to determine if the participants lost weight on the diet, and not just if their weight changed, we want a 1 tailed p-value

Step 4 - Find the *p*-value from the table, and compare to our desired confidence level

We can look up the T value of 1.0757 with a DF of 19 in the T-Table. Once again, the coarseness of the T-table means we cannot get an exact result, but we do see that the result is at least 0.1, which means that we won't have the desired 95% confidence.

One Tail	0.005	0.01	0.025	0.05	0.1	0.25
Two Tail	0.01	0.02	0.05	0.1	0.2	0.5
Degrees of Freedom	Our result is between 0.1 and 0.25					
1	63.6567	31.8205	12.7062	6.3138	3.0777	1.0000
2	9.9248	6.9646	4.3027	2.9200	1.8856	0.8165
3	5.8409	4.5407	3.1824	2.3534	1.6377	0.7649
4	4.6041	3.7469	2.7764	2.1318	1.5332	0.7407
5	4.0321	3.3649	2.5706	2.0150	1.4759	0.7267
6	3.7074	3.1427	2.4469	1.9432	1.4398	0.7176
7	3.4995	2.9980	2.3646	1.8946	1.4149	0.7111
8	3.3554	2.8965	2.3060	1.8595	1.3968	0.7064
9	3.2498	2.8214	2.2622	1.8331	1.3830	0.7027
10	3.1693	2.7638	2.2281	1.8125	1.3722	0.6998
11	3.1058	2.7181	2.2010	1.7959	1.3634	0.6974
12	3.0545	2.6810	2.1788	1.7823	1.3562	0.6955
13	3.0123	2.6503	2.1604	1.7709	1.3502	0.6938
14	2.9768	2.6245	2.1448	1.7613	1.3450	0.6924
15	2.9467	2.6025	2.1314	1.7531	1.3406	0.6912
16	2.9208	2.5835	2.1199	1.7459	1.3368	0.6901
17	2.8982	2.5669	2.1098	1.7396	1.3334	0.6892
18	2.8784	2.5524	2.1009	1.7341	1.3304	0.6884
df 19 → 19	2.8609	2.5395	2.0930	1.7291	1.3277	0.6876
20	2.8453	2.5280	2.0860	1.7247	1.3253	0.6870

1.0757

We can get a more exact p-value using a different source, such as Excel. Looking up a T value of 1.0757 and DF of 19 in excel using = T.Dist.RT() we get a cumulative *p*-value of .1478. This is not less than .05, so we cannot conclude with at least 95% confidence that participants will lose at least 4 pounds.

Putting it all together in one table

Solution By Hand	
Number of samples	21
Mean Change in Weight	-5.35
Weight Change To Test Against	-4.00
Mean Change In Weight less Testing Weight	-1.35
Sample Standard Deviation Change in Weight	5.6127

$$t = \frac{\bar{d} - u_o}{s} * \sqrt{n}$$

T Statistic	-1.102

$$df = n - 1$$

Degrees of Freedom	20

		No, we cannot say with more than 95% confidence that the participants lost more than 4 pounds
Cumulative p value	0.1417	
2 Tailed p value	0.2835	

The 95% confidence t statistic would have been -1.7291.

One Tail	0.005	0.01	0.025	0.05	0.1	0.25
Two Tail	0.01	0.02	0.05	↗ 0.1	0.2	0.5
Degrees of Freedom			Required Confidence Level			
1	63.6567	31.8205	12.7062	6.3138	3.0777	1.0000
2	9.9248	6.9646	4.3027	2.9200	1.8856	0.8165
3	5.8409	4.5407	3.1824	2.3534	1.6377	0.7649
4	4.6041	3.7469	2.7764	2.1318	1.5332	0.7407
5	4.0321	3.3649	2.5706	2.0150	1.4759	0.7267
6	3.7074	3.1427	2.4469	1.9432	1.4398	0.7176
7	3.4995	2.9980	2.3646	1.8946	1.4149	0.7111
8	3.3554	2.8965	2.3060	1.8595	1.3968	0.7064
9	3.2498	2.8214	2.2622	1.8331	1.3830	0.7027
10	3.1693	2.7638	2.2281	1.8125	1.3722	0.6998
11	3.1058	2.7181	2.2010	1.7959	1.3634	0.6974
12	3.0545	2.6810	2.1788	1.7823	1.3562	0.6955
13	3.0123	2.6503	2.1604	1.7709	1.3502	0.6938
14	2.9768	2.6245	2.1448	1.7613	1.3450	0.6924
15	2.9467	2.6025	2.1314	1.7531	1.3406	0.6912
16	2.9208	2.5835	2.1199	1.7459	1.3368	0.6901
17	2.8982	2.5669	2.1098	1.7396	1.3334	0.6892
18	2.8784	2.5524	2.1009	1.7341	1.3304	0.6884
df 19 → 19	2.8609	2.5395	2.0930	1.7291	1.3277	0.6876
20	2.8453	2.5280	2.0860	↗ 1.7247	1.3253	0.6870

Required T Value = 1.7291

If we keep every other number the same and solve for u_0, you can plug that back into the t statistic equation to determine that the 95% confidence weight loss number was 3.18 pounds.

$$t = \frac{\bar{d} - u_o}{s} * \sqrt{n}$$

$$-1.7291 = \frac{-5.35 - u_o}{5.6127} * \sqrt{20}$$

$$u_o = -3.18$$

Alternatively, you could assume that your mean difference and standard deviation results would remain unchanged and calculate how many samples you would need to be 95% confidence that the dieters lost at least 4 lbs. (After all, they lost 5.35 lbs. on average)

$$t = \frac{\bar{d} - u_o}{s} * \sqrt{n}$$

$$-1.7291 = \frac{-5.35 - (-4)}{5.6127} * \sqrt{n}$$

$$n = 51.68$$

The result is that you would need at least 51.68 pairs of samples. Since you cannot have a fraction of a sample, this would be at least 52 samples. (Of course, once you got the additional measurements, you might find that your mean difference of -5.35 lbs. or your standard deviation of 5.6127 lbs. had changed. We also ignored the fact that by getting 52 samples, your degrees of freedom would have changed. However, when we look up the T requirement for 50 degrees of freedom and see that it is 1.68, as opposed to the T requirement of 1.73 for the 19 degrees of freedom, this is not a bad assumption.

Excel Solution for A Paired T-Test with Non-Zero Hypothesis

We can still use this function to do the Paired T-Test in Excel

=TTest(array1, array2, tails, type)

However, since we are not testing against a hypothesized value of zero, there is some editing to the data to do beforehand. The reason we need to edit the data is that there is nowhere to input a hypothesized difference in this function. We need to make sure that hypothesized value is already included in either array1 or array2 of the data that we are feeding the function.

In this example, I am going to do that by subtracting 4 pounds from the before weight before feeding it into the function.

Participant #	Before Weight (lbs)	After Weight (lbs)	Before weight minus 4 lbs
1	123	116	119
2	135	131	131
3	110	102	106
4	115	108	111
5	150	140	146
6	170	172	166
7	108	113	104
8	155	144	151
9	132	139	128
10	118	109	114
11	137	125	133
12	145	140	141
13	181	175	177
14	115	115	111
15	182	173	178
16	145	137	141
17	132	127	128
18	119	116	115
19	127	125	123
20	160	145	156
Average	138.0	132.6	134.0

So, for instance, for the first participant instead of the function comparing a weight of 123 vs 116 lbs. and getting a weight change of 7 lbs., the function will compare 119 vs 116 lbs. and calculate a weight change of 3 lbs., which is performing the same function as subtracting the u_o in our by hand equation

$$t = \frac{\bar{d} - u_O}{s} * \sqrt{n}$$

When you input the modified data into the =TTEST function, you get the same result out as our by hand calculation

86

```
Excel Solution

Paired T Test      0.1478
```

Two-Sample T-Tests

The final two examples are the type of statistical significance calculations you probably most commonly think of for any medicinal scientific study. In these examples, you essentially have two groups, one of which is the control and the other is the study group.

These types of statistical significance calculations are for when you know nothing about the population you are testing. You don't know the mean value or the standard deviation.

Therefore, you have to establish those values for the baseline group as well as for the group you are studying. Unlike the Paired T-Tests we showed in the last example, these tests are done when the members of the population are not the same. I.e. you aren't getting a baseline result from a person and then getting another result from the same person later.

It is useful to visualize what is happening in these calculations. Imagine you own a running shoe company and want to demonstrate that your shoes are faster than your competition. You invite 100 people to your track and time them on a 100-meter run. You provide your shoes to 50 people, and your competitor's shoes to 50 people. (And since you are doing this for science, not marketing, you don't even put weights in your competitor's shoes).

The first result that you will get is the mean values for both groups.

The mean values are instructive, but to get a confidence level you need to know what the probability distribution of the results the mean values are. If the distribution is similar to the chart below

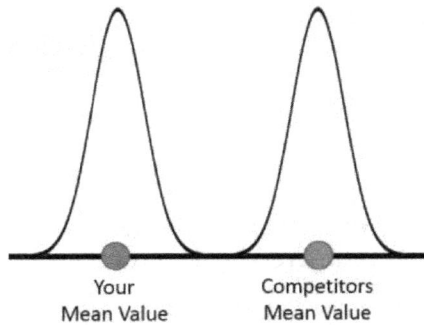

You will be confident that your shoes make people faster. If it the distributions are more like this chart

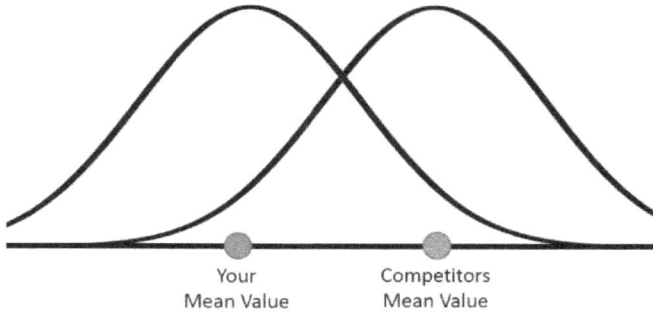

Your Mean Value   Competitors Mean Value

Then your confidence will be a lot lower because, for instance, the true average of your competition's shoe could be somewhere on the left of its standard distribution, and the true average of your shoe could be on the right of yours, as shown below.

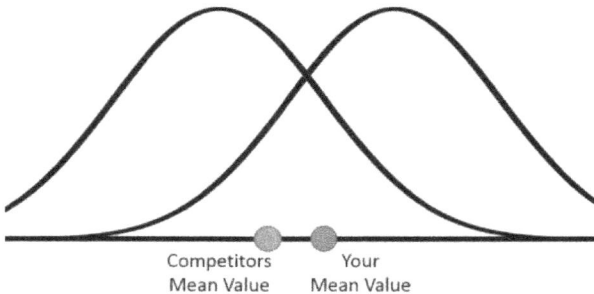

Competitors Mean Value   Your Mean Value

The following two examples show how to do that calculation.

The difference between the two examples is how we are accounting for the standard deviation. Specifically, can we assume that the standard deviation for the two groups is the same? If so, we can just calculate a single standard deviation of all the measurements, as we do in example

4. Otherwise, we need to calculate separate standard deviations for both groups, as we do in example 5.

By pulling 100 people off the street and measuring their running speed, we expect that both groups should be similar in how much difference there is between good runners and bad runners, and therefore it would be applicable to use a single standard deviation.

By contrast, maybe you are trying to determine if there is a difference in fuel economy between cars which come from a brand-new factory where they are entirely built by robots, vs an older factory where they are mostly manually constructed. Not only might the mean values of the two factories differ, it is entirely possible that the new robot factory has a smaller variation in results among its cars. In that case, an unequal variance is probably a better assumption.

Example 4 – Two Sample T-Test with Equal Variance

You have two different types of cat food, and want to determine if the cats eat a different amount of type B food than type A, with at least 95% confidence. You are not using the same cats for each test

Type A, Amount Eaten (ounces)	Type B, Amount Eaten (ounces)
1.4	2.2
2.5	3.0
1.6	1.9
2.3	2.1
0.9	2.7
1.1	1.6
2.7	1.3
2.1	3.0
1.3	3.1
1.4	1.9
	1.7
	1.8

Solving The T-Test with Equal Variance Problem:

Step 1 – Determine What Test to Use

For this problem, we will use a Two Sample T-Test, with equal variance. We do not know the population mean or standard deviation, so we can't do a Z-test. Since we aren't using the same test subjects for the two measurements, we can't do a paired T-Test. Since the measurements are for cats eating food, we are assuming that the variance is close enough between the two samples to use the Equal Variance calculation (which will give a slightly lower p-value than the Unequal Variance calculation) but that is something that we can check while doing the problem.

Note, for this problem we don't have the same number of measurements for Sample A compared to Sample B.

Although the same number of measurements for the two data sets would be required for a Paired T-Test, it is not required for a Two-Sample T-Test, with either equal or unequal variance.

Step 2 – Find the Test Statistic

The equation for t for 2 Sample T-Test with Equal Variance is

$$t = \frac{(\bar{x}_1 - \bar{x}_2)}{\sqrt{\frac{(n_1 - 1)s_1^2 + (n_2 - 1)s_2^2}{n_1 + n_2 - 2}} * \sqrt{\frac{1}{n_1} + \frac{1}{n_2}}}$$

Where:

- $\bar{x}_1$ is the average of Sample 1

- $\bar{x}_2$ is the average of Sample 2

- s_1 is the sample standard deviation of Sample 1

- s_2 is the sample standard deviation of Sample 2

- n_1 is the test sample size of Sample 1

- n_2 is the test sample size of Sample 2

This equation seems more complicated than the previous equations, and it is, but it is understandable if we break it

92

into 3 parts. Those 3 parts are the same 3 parts we have seen in every variant of this equation.

Part 1 - Difference Of Means

$$t = \frac{(\bar{x}_1 - \bar{x}_2)}{\sqrt{\frac{(n_1 - 1)s_1^2 + (n_2 - 1)s_2^2}{n_1 + n_2 - 2}} * \sqrt{\frac{1}{n_1} + \frac{1}{n_2}}}$$

Part 2
Average Standard Deviation

Part 3
Of Samples

Let's dissect this equation

Part 1 – Difference of Means

$$(\bar{x}_1 - \bar{x}_2)$$

is the same difference of mean values that we have seen in every equation so far.

Part 2 – Average Standard Deviation

The

$$\sqrt{\frac{(n_1 - 1)s_1^2 + (n_2 - 1)s_2^2}{n_1 + n_2 - 2}}$$

Looks complicated. On every other equation we have seen, this part of the equation would either just be the population standard deviation σ, or the sample standard deviation s. This equation is different because we have two sample standard deviations s_1 and s_2 from our two sets of data. However, remember that theoretically s_1 and s_2 are equal because this is the equal variance problem. Variance is standard deviation squared.

So, what do you do when you have two values, but need one value for your equation? Take the average of them, or in this case, the weighted average.

The sample variance of set 1 uses (n_1-1) in its equation

$$s_1^2 = \frac{\Sigma(x_1 - \overline{x_1})^2}{\boxed{n_1 - 1}}$$

The variance of set 2 uses (n_2-1) in its equation. The (n_1-1) and (n_2-1) can be multiplied by their respective variances to get a weighted standard deviation. And since we need a weighted <u>average</u>, we then divide by $(n_1-1) + (n_2-1)$, which is the $(n_1 + n_2-2)$ part of the equation.

$$\sqrt{\frac{\overbrace{(n_1 - 1)}^{\text{Weighted}} s_1^2 + (n_2 - 1) s_2^2}{\underbrace{n_1 + n_2 - 2}_{\text{Average}}}}$$

Two Variances

The result is a single variance.

Part 3 – Number of Samples

Now what about the

$$\sqrt{\frac{1}{n_1} + \frac{1}{n_2}}$$

part of the equation?

This is taking the place of the

$$\sqrt{n}$$

that we initially saw in the average of dice rolls example, and have seen in all the other equation also. Once again, more measurements in the form of either more n_1 or more n_2 will increase the t value.

95

An Example of How That Equation Works

When we looked at rolling the dice, we found that the standard deviation of the mean was captured by the equation

$$\frac{\sigma}{\sqrt{n}}$$

But what we care about here is the standard deviation of the distance between two means (with equal variances). This is similar to saying that we are rolling a red die and a blue die. Both have 6 sides. For the red die let's keep its values as is. For the blue die let's add 2 to its roll. How does the variance between mean values change?

It turns out that whichever data set has the smallest number of samples (the smaller n) tends to dominate the result. Here is a chart with 1 roll of both dice

Standard Deviation Of The Mean
For Both Dice = 1.708

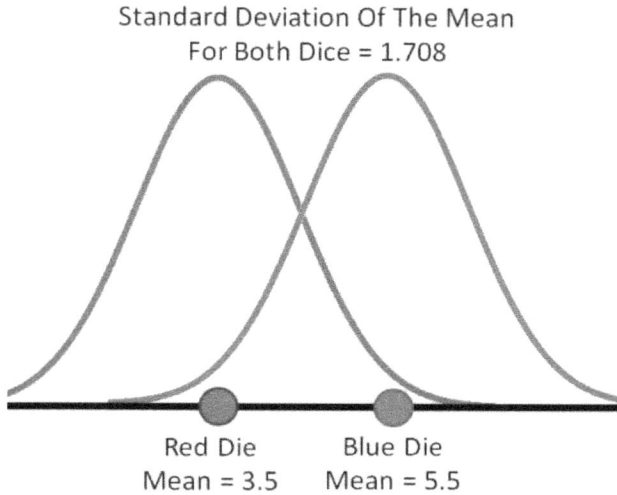

Red Die   Blue Die
Mean = 3.5   Mean = 5.5

Now let's say you roll the blue die 20 times and average its value and plot those results over many sets of the 20 rolls. Its standard deviation of the average of the mean will drop dramatically, and the average of the mean of the difference will be dominated by the red die.

Standard Deviation Of The Mean
Red Die = 1.708
Blue Die = 1.708 / √20 = .382

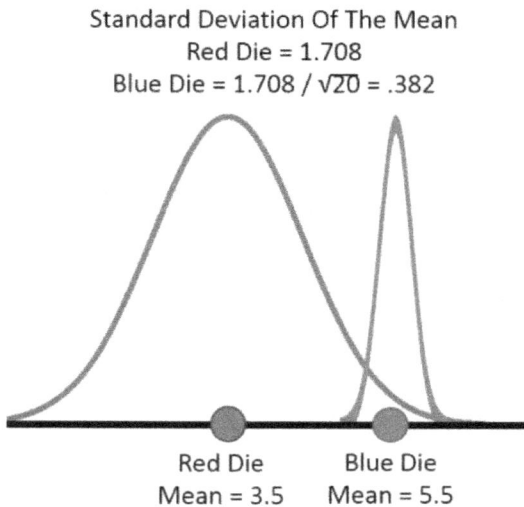

Red Die   Blue Die
Mean = 3.5   Mean = 5.5

Remember, these two values have the same standard deviation of values, which is 1.708, but the standard deviation of their mean is driven by the equation

$$\frac{\sigma}{\sqrt{n}}$$

And in the chart above we have only rolled the red die 1 time, but we have rolled the blue die 20 times. As a result, we have more certainty on what the mean value we expect from the blue die is.

What Would Happen If You Rolled One Die Many, Many Times?

If I roll the blue die many more times, say infinitely many times, what ends up happening is that the count of n_2 and the variance of s_2 drop out of the equation

$$\sqrt{\frac{1}{n_1} + \frac{1}{\infty}} = \sqrt{\frac{1}{n_1}}$$

And a sample deviation with infinite samples just becomes a population deviation

$$\sqrt{\frac{(n_1 - 1)s_1^2 + (\infty - 1)\sigma^2}{n_1 + \infty - 2}} = \sqrt{\frac{\infty\sigma^2}{\infty}} = \sigma$$

And what we are left with is the

$$t = \frac{(\bar{x}_1 - \bar{x}_2)}{\dfrac{\sigma}{\sqrt{n_1}}}$$

This is the same equation as we had in our first example. Getting data for just one half of the problem can't get rid of all the uncertainty, to do that you need to address both sets of data. Visually what happened is if we gather enough data for the blue die, eventually the standard deviation of the mean becomes infinitely small, and the blue curve will be infinitely narrow and we will get this

Standard Deviation Of The Mean
Red Die = 1.708
Blue Die = 0

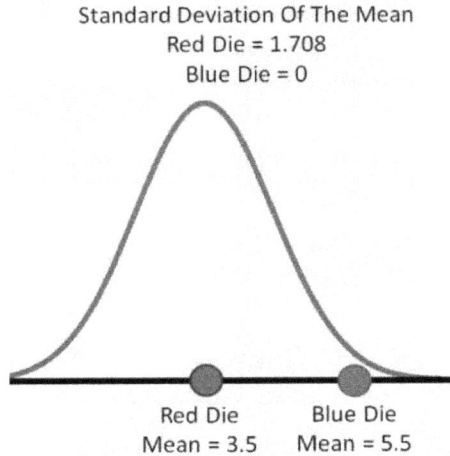

Red Die   Blue Die
Mean = 3.5   Mean = 5.5

The t values for this type of problem will tend to be lower than the t values we would have for the equations we used in Examples 1 or 2, when we knew the mean value of what we were comparing against for certain. The reason for this is clear when you consider that the Z equation in example 1 is just a specific instance of this equation, an instance where you have infinitely many samples for one of your data sets so you know the population mean and standard deviation.

Example 4 – Two Sample with Equal Variance - Solution:

It is simple to use these functions in Excel to find Average, Sample Standard Deviation, and number of samples of the data. You can also see Example 3 for an example of finding sample standard deviation by hand

=Average() for average

=STDEVA() for sample standard deviation (note the difference between that and population standard deviation)
=Count() for sample size

	Type A, Amount Eaten (ounces)	Type B, Amount Eaten (ounces)
	1.4	2.2
	2.5	3.0
	1.6	1.9
	2.3	2.1
	0.9	2.7
	1.1	1.6
	2.7	1.3
	2.1	3.0
	1.3	3.1
	1.4	1.9
		1.7
		1.8
Average	1.73	2.19
Sample Standard Deviation	0.624	0.611
Number of Measurements	10	12

Here we see that the two standard deviations are .624 and .611. Those values are close enough that the equal variances assumption is a good one. When we plug those values into our t equation we get

$$t = \frac{(\bar{x}_1 - \bar{x}_2)}{\sqrt{\frac{(n_1 - 1)s_1^2 + (n_2 - 1)s_2^2}{n_1 + n_2 - 2}} * \sqrt{\frac{1}{n_1} + \frac{1}{n_2}}}$$

$$t = \frac{(1.73 - 2.19)}{\sqrt{\frac{(10 - 1).624^2 + (12 - 1).611^2}{10 + 12 - 2}} * \sqrt{\frac{1}{10} + \frac{1}{12}}}$$

$$t = \frac{-0.46}{.6168 * .428} = -1.748$$

With the summarized information, the t statistic is calculated to be -1.748

Degrees of Freedom

The formula for degrees of freedom for Equal Variance is

$$df = n_1 + n_2 - 2$$

This formula comes from adding the $(n_1\text{-}1)$ and $(n_2\text{-}1)$ that we would use to get the degrees of freedom if the two variances of the mean were separate problems.

So, with

$n_1 = 10$ and $n_2 = 12$ we have 20 degrees of freedom

Step 3 - 1 Tailed or 2 Tailed

Since we want to determine if the cats ate a different amount of food, as opposed to just more food, we want a 2-tailed *p*-value

Step 4 - Find the *p*-value from the table, and compare to our desired confidence level

When we look up the value of 1.748 in a T table, with DF=20 we can find the value of 1.7247 which is pretty close. That gives a p value of .1 for the 2-tailed test.

One Tail	0.005	0.01	0.025	0.05	0.1	0.25
Two Tail	0.01	0.02	0.05	0.1	0.2	0.5
Degrees of Freedom						
1	63.6567	31.8205	12.7062	6.3138	3.0777	1.0000
2	9.9248	6.9646	4.3027	2.9200	1.8856	0.8165
3	5.8409	4.5407	3.1824	2.3534	1.6377	0.7649
4	4.6041	3.7469	2.7764	2.1318	1.5332	0.7407
5	4.0321	3.3649	2.5706	2.0150	1.4759	0.7267
6	3.7074	3.1427	2.4469	1.9432	1.4398	0.7176
7	3.4995	2.9980	2.3646	1.8946	1.4149	0.7111
8	3.3554	2.8965	2.3060	1.8595	1.3968	0.7064
9	3.2498	2.8214	2.2622	1.8331	1.3830	0.7027
10	3.1693	2.7638	2.2281	1.8125	1.3722	0.6998
11	3.1058	2.7181	2.2010	1.7959	1.3634	0.6974
12	3.0545	2.6810	2.1788	1.7823	1.3562	0.6955
13	3.0123	2.6503	2.1604	1.7709	1.3502	0.6938
14	2.9768	2.6245	2.1448	1.7613	1.3450	0.6924
15	2.9467	2.6025	2.1314	1.7531	1.3406	0.6912
16	2.9208	2.5835	2.1199	1.7459	1.3368	0.6901
17	2.8982	2.5669	2.1098	1.7396	1.3334	0.6892
18	2.8784	2.5524	2.1009	1.7341	1.3304	0.6884
19	2.8609	2.5395	2.0930	1.7291	1.3277	0.6876
DF 20 → 20	2.8453	2.5280	2.0860	1.7247	1.3253	0.6870

Looking up a T value of 1.748 and DF of 20 in excel using = T.Dist.2T() we get a cumulative *p*-value of .0958. This is not less than .05, so we cannot conclude with at least 95% confidence that cats will eat a different amount of food. (Note, the 1 tailed *p*-value is .0479, so we could have concluded with 95% confidence that the cats eat more food), which is an odd, but correct, result and shows the importance of deciding if you intend to get a 1 tailed or 2 tailed result.

Summarizing this into one table

Solution By Hand	
Average, Sample 1	1.730
Sample Standard Deviation, Sample 1	0.624
Number of Measurements, Sample 1	10
Average, Sample 2	2.192
Sample Standard Deviation, Sample 2	0.611
Number of Measurements, Sample 2	12

$$t = \frac{(\bar{x}_1 - \bar{x}_2)}{\sqrt{\frac{(n_1 - 1)s_1^2 + (n_2 - 1)s_2^2}{n_1 + n_2 - 2}} * \sqrt{\frac{1}{n_1} + \frac{1}{n_2}}}$$

T Statistic	-1.748

$$df = n_1 + n_2 - 2$$

Degrees of Freedom	20	
Cumulative *p* value	0.0479	
2 Tailed *p* value	0.0958	No, we cannot say with more than 95% confidence that the cats eat a different amount of type B food

Excel Solution:

The function to do this problem in Excel is

=TTEST(array1, array2, 2, 2)

Where the 2 that is entered for the third piece of information is because it is a 2-tailed problem, and the 2 that is entered for the fourth piece of information selects the 2 Sample, Equal Variance solution type

Excel Solution	
2 Sample T Test, Equal Variance	0.0958

Example 5 - 2 Sample T-Test with Unequal Variance

This problem goes over the last type of statistical significance. It is two sets of data that do not have the same variance.

This leads to the question, when will you have data that you want to compare which you cannot assume has equal variance? For a lot of examples, equal variance is a decent assumption. If I pull 20 people off the street and give them either a standard golf club or my new fancy golf club, I might see that they can hit the ball farther with the new golf club, but there will probably be similar scatter among the members in the two groups.

Of course, one would expect different variance if the two sets of data were quite dissimilar, for instance, the average weight of a group of dogs vs a group of coyotes. But what is an example for when you have two similar sets of data? When can you expect to get an unequal variance?

The best example I have is if you have introduced some quality control into a process. For instance, if you measure the average amount of time it takes some workers to perform a task, you might get one level of variance. If you then introduce training so that everyone does the job the same way instead of their own way, you can expect less scatter in how long it takes people to perform that task, in addition to, hopefully, an overall reduction in average time.

Things that introduce training or other quality control are likely to change the standard deviation of the two sets of data and make it a good time to use this equation with unequal variance.

Of course, you can calculate the variance of your data sets and determine how similar they are

In terms of the equation itself, it is largely similar to all the other variants that we have seen so far, and most of the process is exactly the same.

Unequal Variance Example Problem

You own a factory, and you have measured the time it takes some of your employees to assemble a part. You then give a different group of employees training on how to do the job better, and measure how long it takes them to assemble the part. You want to determine, with 90% confidence, if the training has reduced the average time it takes to do that job.

Average Time Before Training (minutes)	Average Time After Training (minutes)
10	8
19	8
17	9
14	10
20	11
9	13
6	15
15	14
19	15
5	13
7	12
20	13
14	13
15	
19	

Solving The 2 Sample Unequal Variance Problem:

Step 1 – Determine What Test to Use

For this problem, we will use a Two Sample T-Test, with unequal variance. We do not know the population mean or standard deviation, so we can't use the Z-test. Since we aren't using the same test subjects for the two measurements, we can't do a paired T-Test. The question becomes "Should you use the 2 sample T test equation with equal variance or unequal variance?"

In this case, we are assuming that we will need to use the equation for unequal variance since your training very well could have significantly reduced the variance in the amount of time it takes different people to do the job. In fact, when we calculate the sample variance for the two sets, we find that the sample standard deviation for the first set is 5.298 minutes, and the sample standard deviation for the second set is 2.444 minutes. Those results are different enough that it makes sense to use the equation for unequal variance.

Step 2 – Find the Test Statistic

The equation for t for 2 Sample T-Test with Unequal Variance is

$$t = \frac{(\overline{x_1} - \overline{x_2})}{\sqrt{\frac{s_1^2}{n_1} + \frac{s_2^2}{n_2}}}$$

Where:

- $\overline{x}_1$ is the average of Sample 1

- $\overline{x}_2$ is the average of Sample 2

- s_1 is the sample standard deviation of Sample 1

- s_2 is the sample standard deviation of Sample 2

- n_1 is the test sample size of Sample 1

- n_2 is the test sample size of Sample 2

Dissecting the equation

$$(\overline{x_1} - \overline{x_2})$$

is the same difference of means that we see in every problem. For the standard deviation part of the equation

$$\sqrt{\frac{s_1^2}{n_1} + \frac{s_2^2}{n_2}}$$

We see a similar effect as we do in Example 4. The larger of the s^2/n terms is going to dominate the problem. If you have one of those terms that is really large (due to having a high sample variance or a really small number of samples) then no matter how small the other term is you will be limited in the resulting t statistic.

If you get a large enough number of measurements, either n_1 or n_2, then one of the terms will drop out of the problem, and the resulting equation will be the same as we have in example 2. Here is how the equation would look if n_2 went to infinity.

$$t = \frac{(\overline{x_1}-\overline{x_2})}{\sqrt{\frac{s_1^2}{n_1}+\frac{s_2^2}{\infty}}} = \frac{(\overline{x_1}-\overline{x_2})}{\sqrt{\frac{s_1^2}{n_1}}}$$

$$t = \frac{(\overline{x_1}-\overline{x_2})}{\frac{s_1}{\sqrt{n_1}}}$$

"Large enough" is, of course, a relative term, but for general engineering, if you have an s_2/n term that is at least two orders of magnitude smaller than the other variance of the mean term, then the smaller value will tend to have very little impact on the results. (You could probably get away with ignoring something that is at least 1 order of magnitude smaller than the other s^2/n term, depending on what you are doing the data analysis for)

It is simple to use these functions in Excel to find Average, Sample Standard Deviation, and a number of samples. You can also see Example 3 for an example of finding sample standard deviation by hand

=Average() for average
=STDEVA() for sample standard deviation (note the difference between that and population standard deviation)
=Count()   for sample size

	Average Time Before Training (minutes)	Average Time After Training (minutes)
	10	8
	19	8
	17	9
	14	10
	20	11
	9	13
	6	15
	15	14
	19	15
	5	13
	7	12
	20	13
	14	13
	15	
	19	
Average	13.93	11.85
Sample Standard Deviation	5.298	2.444
Number of Measurements	15	13

Get the Data & Excel Functions

If you want the excel file that contains all of the example data and solution, it can be downloaded for Free here http://www.fairlynerdy.com/statistical-significance-examples/

The T-Statistic equation is

111

$$t = \frac{(\overline{x_1} - \overline{x_2})}{\sqrt{\frac{s_1^2}{n_1} + \frac{s_2^2}{n_2}}}$$

Plugging our values into the equation, the t statistic is calculated to be 1.36

$$t = \frac{(13.93 - 11.85)}{\sqrt{\frac{5.298^2}{15} + \frac{2.444^2}{13}}} = 1.36$$

Degrees of Freedom

The formula for degrees of freedom for Unequal Variance is fairly complicated, it is

$$df = \frac{\left(\frac{s_1^2}{n_1} + \frac{s_2^2}{n_2}\right)^2}{\frac{\left(\frac{s_1^2}{n_1}\right)^2}{n_1 - 1} + \frac{\left(\frac{s_2^2}{n_2}\right)^2}{n_2 - 1}}$$

So, with plugging in the values from our data

$$df = \frac{(\frac{5.298^2}{15} + \frac{2.444^2}{13})^2}{\frac{(\frac{5.298^2}{15})^2}{15-1} + \frac{(\frac{2.444^2}{13})^2}{13-1}} = 20.3$$

We have 20.3 degrees of freedom, which can just be rounded to 20

This degree of freedom equation is another case where the result is primarily driven by the smaller of n_1 and n_2. In this case, the two values are similar in magnitude, but if one was significantly larger than the other, the smaller value would end up controlling the resulting degrees of freedom. The chart below shows what the degree of freedom value would be, using the same 5.298 and 2.444 values for variance as we did above, if we had a range of different n_1 and n_2 values

Degrees of Freedom vs n_1 & n_2										
$n_1 \backslash n_2$	1	2	3	4	5	6	7	8	9	10
1										
2		1.4	1.3	1.2	1.2	1.1	1.1	1.1	1.1	1.1
3		2.9	2.8	2.6	2.5	2.4	2.4	2.3	2.3	2.3
4		4.0	4.4	4.2	4.0	3.9	3.7	3.7	3.6	3.5
5		4.4	5.9	5.9	5.6	5.4	5.2	5.1	5.0	4.9
6		4.4	7.0	7.4	7.3	7.0	6.8	6.6	6.4	6.3
7		4.2	7.7	8.8	8.9	8.7	8.4	8.2	8.0	7.8
8		4.0	8.1	10.0	10.5	10.4	10.1	9.9	9.6	9.4
9		3.7	8.2	10.9	11.8	12.0	11.8	11.5	11.3	11.0
10		3.4	8.1	11.4	13.0	13.5	13.4	13.2	12.9	12.7
11		3.2	7.8	11.7	13.9	14.8	15.0	14.9	14.6	14.4
12		3.0	7.6	11.8	14.6	16.0	16.5	16.5	16.3	16.1
13		2.8	7.3	11.8	15.1	17.0	17.8	18.0	18.0	17.7
14		2.7	7.0	11.6	15.4	17.7	19.0	19.5	19.6	19.4
15		2.6	6.7	11.4	15.5	18.3	20.0	20.8	21.1	21.0
16		2.5	6.4	11.1	15.5	18.7	20.8	22.0	22.5	22.6
17		2.4	6.2	10.8	15.4	19.0	21.5	23.0	23.8	24.1
18		2.3	5.9	10.5	15.2	19.1	22.0	23.9	24.9	25.5
19		2.2	5.7	10.2	14.9	19.1	22.4	24.6	26.0	26.8
20		2.1	5.5	9.9	14.7	19.1	22.6	25.2	26.9	27.9

As you can see, by looking down the $n_2 = 2$ column for instance, as n_1 gets very large the degree of freedom values is driving by n_2.

If n_1 or n_2 get large enough to be considered infinity (once again, you can estimate that as two orders of magnitude larger than the other term) then it cancels out of the equation and you are left with a much simpler equation.

For instance, if n_2 approaches infinity, the equation becomes

$$df = \frac{(\frac{s_1^2}{n_1} + \cancel{\frac{s_2^2}{\infty}})^2}{\frac{(\frac{s_1^2}{n_1})^2}{n_1 - 1} + \cancel{\frac{(\frac{s_2^2}{\infty})^2}{\infty - 1}}}$$

This ends up reducing to

$$df = \frac{\left(\frac{s_1^2}{n_1}\right)^2}{\frac{\left(\frac{s_1^2}{n_1}\right)^2}{(n_1-1)}} = \frac{\left(\frac{s_1^2}{n_1}\right)^2}{\left(\frac{s_1^2}{n_1}\right)^2}(n_1 - 1)$$

And finally simplifies to

$$df = (n_1 - 1)$$

Which of course is the same equation for degrees of freedom that we saw in Example #2.

Step 3 - 1 Tailed or 2 Tailed

Since we want to determine if the workers do the job in less time, not just a different amount of time, we will use a 1 tailed p-value

Step 4 - Find the p-value from the table, and compare to our desired confidence level

Looking up a T value of 1.367 and DF of 20.3 in excel using = T.Dist.RT() we get a cumulative p-value of .0934. This is less than .1, which means we have a greater than 90% confidence that the mean value of the first data set is greater than the mean value of the second data set, i.e. the training did cause a decrease in the average time it took the workers to assemble the parts.

If we were to look this result up in the same T-table we have been using, we would run into the interesting problem that a DF value of 20.3 is not in the table. (In addition to the T value of 1.367 not being directly in the table.) As a result, we could use the closest value in the table, which would be DF of 20, or we could interpolate.

Summarizing this into one table

	Solution By Hand	
Number of Measurements, Sample 1	15	
Number of Measurements, sample 2	13	

$$t = \frac{(\bar{x}_1 - \bar{x}_2)}{\sqrt{\dfrac{s_1^2}{n_1} + \dfrac{s_2^2}{n_2}}}$$

T Statistic	1.367

$$df = \frac{\left(\dfrac{s_1^2}{n_1} + \dfrac{s_2^2}{n_2}\right)^2}{\dfrac{\left(\dfrac{s_1^2}{n_1}\right)^2}{n_1 - 1} + \dfrac{\left(\dfrac{s_2^2}{n_2}\right)^2}{n_2 - 1}}$$

Degrees of Freedom	20.29

Yes, we have a greater than 90% confidence that the mean value of set 1, before training, is greater than the mean value of set 2, after training

Cumulative p value	0.0934
2 Tailed p value	0.1867

Get the Data & Excel Functions

If you want the excel file that contains all of the example data and solution, it can be downloaded for Free here http://www.fairlynerdy.com/statistical-significance-examples/

Excel Solution:

The function to do this problem in Excel is

=TTEST(array1, array2, 1, 3)

Where the 1 that is entered for the third piece of information is because it is a 1 tailed problem, and the 3 that is entered for the fourth piece of information selects the 2 Sample, Unequal Variance solution type.

Once again, we get a 1 tailed confidence of .0933, this means we have a greater than 90% confidence that the workers from the first data set took longer than the workers from the second data set to assemble the parts.

Excel Solution	
2 Sample T Test, Equal Variance	0.0933

Chapter 6: Different Types of Variances

The last two problems were quite similar. The only difference between them is if we considered the data to have equal variance or unequal variance when we chose our equations. That, of course, leaves some room for a judgment call. So, what would our results have been if we had done the problem the other way? For instance, what if we had used the data in Example 4 and assumed unequal variance, or if we had used the data in Example 5 but assumed they had equal variance?

Here is what the effect would have been on the test statistic, the degrees of freedom, and the resulting 1 tailed confidence. (Note, the original example 4 used 2 tailed confidence, but this table uses 1 tailed to better match example 5)

Here are the results for Example 4 data

Example 4		
	Equal Variance	Unequal Variance
t	-1.748	-1.745
df	20	19.140
1 Tailed p	0.0479	0.0486

Here are the results for Example 5 data

Example 5		
	Equal Variance	Unequal Variance
t	1.303	1.367
df	26	20.293
1 Tailed p	0.1020	0.0934

Based on these results we cannot categorically say that either the equal or unequal variance equation will give a higher confidence level or t value. It will depend on the data. The Equal Variance equation will always give a higher degree of freedom. However, as we saw, the importance of the degree of freedom drops of quickly once you get more than 10 samples or so.

For practical purposes, however, the results between the two equations are not that different. If you are not sure whether equal or unequal variance is the right solution for your problem, you can solve it both ways and see the results, or just pick one and go with it.

Last Chance to Get YOUR Bonus!

FOR A LIMITED TIME ONLY – Get the best-selling book *"5 Steps to Learn Absolutely Anything in as Little As 3 Days!"* by Edward Mize absolutely FREE!

Readers who have read this bonus book as well have seen huge increases in their abilities to learn new things and apply it to their lives – so it is *highly recommended* to get this bonus book.

Once again, as a big thank-you for downloading this book, I'd like to offer it to you *100% FREE for a LIMITED TIME ONLY!*

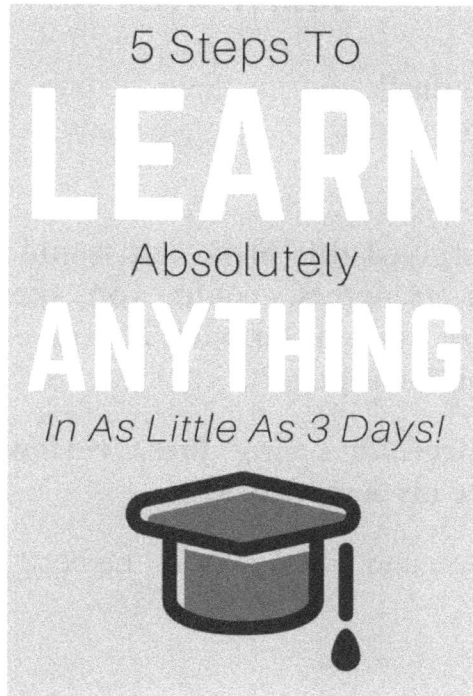

TEACHINGNERDS.COM

5 Steps To
LEARN
Absolutely
ANYTHING
In As Little As 3 Days!

To download your FREE copy, go to:

TeachingNerds.com/Bonus

Final Words

I would like to thank you for downloading my book and I hope I have been able to help you and educate you on something new.

If you have enjoyed this book and would like to share your positive thoughts, could you please take 30 seconds of your time to go back and give me a review on my Amazon book page!

I greatly appreciate seeing these reviews because it helps me share my hard work!

Again, thank you and I wish you all the best!

Disclaimer

This book and related sites provide information in an informative and educational manner only, with information that is general in nature and that is not specific to you, the reader. The contents of this site are intended to assist you and other readers in your education efforts. Consult an expert regarding the applicability of any information provided in our books and sites to you.

Nothing in this book should be construed as personal advice, legal advice, or expert advice, and must not be used in this manner. The information provided is general in nature. This information does not cover all possible uses, actions, precautions, consequences, etc. such as loss of data or hardware failure.

You should consult with an expert before applying anything in this book. This book should not be used in place of learning from a professional or seeking advice from a technical specialist.

No Warranties: The authors and publishers don't guarantee or warrant the quality, accuracy, completeness, timeliness, appropriateness or suitability of the information in this book, or of any product or services referenced by this book, other books, and websites.

The information in this book and on relevant websites is provided on an "as is" basis and the authors and publishers make no representations or warranties of any kind with respect to this information. This site may contain inaccuracies, typographical errors, or other errors.

www.ingramcontent.com/pod-product-compliance
Lightning Source LLC
Chambersburg PA
CBHW030528210326
41597CB00013B/1063